SOVIET MEDICAL REVIEWS

A series of journals publishing the most significant developments in the important research areas of Soviet medicine. In making these publications available, the publisher hopes to contribute to the further development of international cooperation between scholars and to a greater understanding among physicians and scientists.

VIROLOGY REVIEWS
edited by D.K. Lvov, D.I. Ivanovskii Institute of Virology, USSR Academy of Medical Sciences, Moscow.

Section E of Soviet Medical Reviews

GENERAL INFORMATION

Aims and Scope. VIROLOGY REVIEWS publishes review articles covering significant developments in Soviet fundamental and applied virology research as well as plans for future experiments. It is intended to make accounts of recent scientific advances in the USSR readily and rapidly available to Western scientists who do not read Russian.

First published 1990 by Harwood Academic Publishers

Published 2022 by Routledge
2 Park Square, Milton Park, Abingdon, Oxon OX14 4RN
605 Third Avenue, New York, NY 10017

Routledge is an imprint of the Taylor & Francis Group, an informa business

Copyright © 1990 by Taylor & Francis

All rights reserved. No part of this book may be reprinted or reproduced or utilised in any form or by any electronic, mechanical, or other means, now known or hereafter invented, including photocopying and recording, or in any information storage or retrieval system, without permission in writing from the publishers.

Notice:
Product or corporate names may be trademarks or registered trademarks, and are used only for identification and explanation without intent to infringe.

Disclaimer
The publisher has made every effort to trace copyright holders and welcomes correspondence from those they have been unable to contact.

ISBN 13: 978-3-7186-4942-6 (pbk)

AF802792

SOVIET MEDICAL REVIEWS

An imprint under which are published authoritative reviews of the most significant developments in important research areas of Soviet medical science.

Section A: Cardiology Reviews edited by E.I. Chazov and V.N. Smirnov

Section B: Physicochemical Aspects of Medicine Reviews edited by Yu. M. Lopukhin

Section C: Hematology Reviews edited by O.K. Gavrilov

Section D: Immunology Reviews edited by R.V. Petrov

Section E: Virology Reviews edited by D.K. Lvov

Section F: Oncology Reviews edited by N.N. Blokhin and N.N. Trapeznikov

Section G: Neuropharmacology Reviews edited by A. V. Valdman

INFLUENZA

CONTENTS

FOREWORD

Volume 3 Part 3 of Virology Reviews concentrates on the problem of influenza, individual aspects of which have already been discussed in several of the papers in Volumes 1 and 2.

Some of the immunological mechanisms of infection, the fate of a viral population in the process of persisting infection, and the possible epidemiological implications of this phenomenon are dealt with here.

Information is also presented about the etiological pattern of the group of acute respiratory diseases in the USSR, the epidemiology of influenza and its prevention.

Antigenic drift in the hemagglutinin of the strains of the Hong Kong (H3N2) virus over the entire period of its circulation is reviewed, and the final article describes the results of the radioimmunoassay of the viral nucleoprotein. This volume does not, of course, exhaust research on this globally important problem; we shall be returning to it in future publications.

D.K. Lvov

Sov. Med. Rev. E. Virology Reviews, Vol. 3. 1990, pp. 1–18
Photocopying permitted by license only

THE IMMUNOLOGY OF INFLUENZA

Ya. S. SHVARTSMAN

All-Union Research Institute of Influenza, USSR Ministry of Health, Leningrad

Abstract

Anti-influenza immunity is a complicated phenomenon: its components include respiratory tract and blood serum antibodies, and specific sensitized lymphocytes. In protecting the organism immunity factors pronouncedly cooperate with each other as well as with factors of non-specific resistance, such as interferon, macrophages and natural killer cells. Apart from preventing infection and the development of disease, immunity mechanisms are also involved in the organism's recovery, the prevention of virus transmission from infected to healthy people, and in selecting drift variants of influenza virus A. The variability of antigen specificity of surface glycoproteins, means that anti-influenza immunity acts, first of all, against definite virus strains and sub-types. The existence of species-specific, "universal" immunity as well, opens up prospects for developing fundamentally new methods of influenza prophylaxis.

Key words: influenza; antibodies; cell-mediated immunity; universal vaccine

1. INTRODUCTION

Immunity is the key problem in the study of influenza. Immunity protects against infection and disease; raising the level of immunity in a population as a whole results in the halting of epidemics. Knowledge about immunity underlies the specific prophylaxis of influenza and further study of the mechanisms and the distinctive patterns of immunity, will allow the development of fundamentally new "universal" vaccines, thus turning influenza into a controllable infection.

2. ANTIBODIES IN SECRETIONS OF THE RESPIRATORY TRACT

In the 1940s it was established that the immunity of mice to influenza infection correlated better with the concentration of neutralizing antibodies in the respiratory tract secretions than with those in serum. Twenty years later it was found that the bulk of secretory antibodies belonged to a special form of immunoglobulin-A(SIgA), which is contained in trace amounts in blood serum. The major biological peculiarities of SIgA antibodies include: higher resistance to proteolytic enzymes than in isotype G antibodies, the ability to neutralize the infectious activity of viruses, and to prevent their attachment to cells; the antibodies can firmly attach to the surface membrane of the cells of respiratory tract epithelium (Shvartsman *et al.*, 1978). Upon interaction with antigens, secretory antibodies do not activate complement through the classical mechanism. Both lymphocytes and macrophages lack receptors for the Fc fragments of these antibodies. According to some authors, the specific activity spectrum of secretory antibodies is broader than that of serum antibodies. SIgA antibodies are synthesized by lymphoid cells localized in mucosal tissue proper (see for review Shvartsman & Zykov, 1976; Shvartsman *et al.*, 1985). The discovery that the organism has a unified local immunity system is a recent breakthrough (Dhar & Ogra, 1985). This system is based on lymphoid tissue associated with the bronchi, intestines, mammary glands, etc. Areas of mucosa where lymphoid cells are localized are covered with specialized epithelial M cells. It is here that the antigens, delivered to the T and B lymphocytes by macrophages, are absorbed. These lymphocytes have an ability to "home in" on mucosal lymphoid aggregates and recirculate between them — a fact which may explain, for instance, the higher concentration of antibodies to influenza A virus in colostrum samples taken from lying-in women in the first few months after an epidemic outbreak (Figure 1).

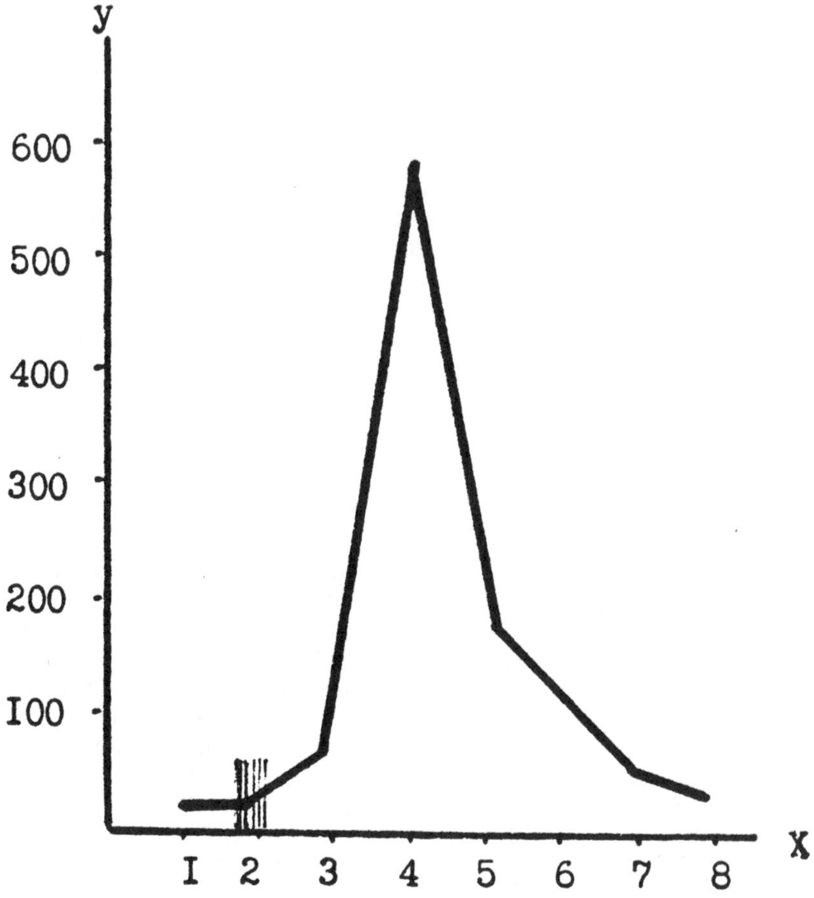

Figure 1 Titres of antibodies to influenza virus A(H3N2) in colostrum samples collected from healthy mothers: *ordinate*, geometric mean antibody titres; *abissa*, month when colostrum samples were collected. The shaded section represents the period of the influenza epidemic.

In the case of natural infection and immunization with a live vaccine, the latent period of antibody production is from 3 to 14 days. Their maximum concentration in the upper respiratory tract is usually observed during the same period. In people who have suffered severe forms of influenza, the titres of secretory antibodies are significantly higher than those of convalescents from mild forms of influenza. The period of retention of secretory antibodies is considerably shorter than that of serum antibodies.

In convalescents with a past history of severe forms of influenza, detectable concentrations of secretory antibodies were retained longer than in people who had influenza with mild clinical manifestations (Shvartsman & Grigorieva, 1979).

Experiments and observations conducted in the laboratory where the author works, have confirmed the availability of an immunological memory in the system of secretory antibody synthesis. With rats immunized intranasally, the optimum period for development of the secondary immune reaction was 45 days. The "original antigenic sin" phenomenon in the local immunity system was also observed in experiments with these animals (Angelova & Shvartsman, 1982). Reimmunization 3 weeks later with a live human vaccine was accompanied by a significant increase in secretory antibody production. Secretory antibody production was more active following the application of antigens directly to the mucosal surface; a relatively autonomous production of secretory and serum antibodies has been described. Both these patterns are statistically significant. If large amounts of virus are administered parenterally, then the secretory antibodies of isotype A are intensively synthesized, due to the generalization of antigen stimulation. Intranasal immunization of mice with an inactivated vaccine with added adjuvant, triggers intensive formation of both secretory and serum antibodies (Shvartsman *et al.*, 1982).

The role that secretory antibodies play in immune protection is still an open question. Two approaches have been tried in our laboratory. The correlation between the levels of the two types of antibody and morbidity was investigated during outbreaks of influenza A and B (Shvartsman *et al.*, 1977; Tsibulskaya *et al.*, 1983). Mice were given a preparation of secretory antibodies intranasally and were then infected with influenza virus. Initial studies included the collection of serum samples and nasopharyngeal washings before and after the influenza outbreaks; for children who came down with the disease such samples were collected on the first or second day of illness and again 2–3 weeks later. Some 300 schoolchildren were kept under observation.

For secretory antibodies the protective titre was $\geq$ 1:8 (by the indirect hemagglutination test), and for serum antibodies was $\geq$ 1:40 (by the hemagglutination-inhibition (HI) test); these titres prevented development of disease. Cooperation between secretory and serum antibodies in protecting the organism could clearly be seen (Figure 2).

During the influenza B outbreak, practically 100% of all boarding-school children under surveillance were infected, but in the majority of them a significant shift in the level of secretory antibodies was recorded. This can be regarded as evidence of the involvement of secretory antibodies in halting the epidemic.

In the second investigation approach, a fraction containing SIgA antibodies was isolated from colostrum samples collected two months

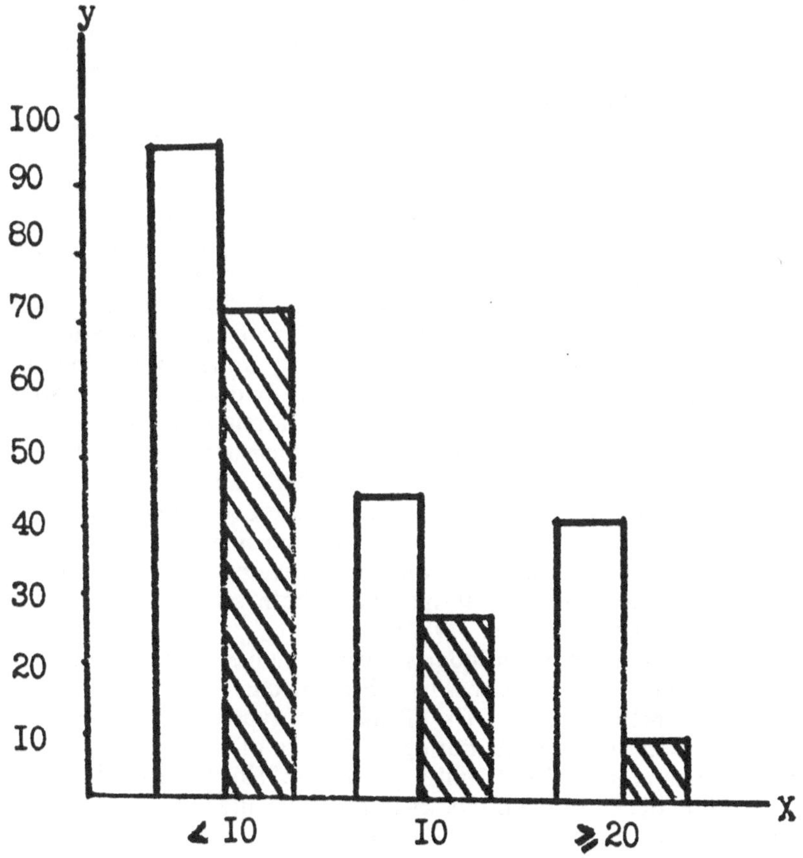

Figure 2 Cooperation between secretory antibodies and those of blood serum in influenza B protection: *ordinate*, disease incidence (%); *absissa*, reciprocal titres of antibodies in blood serum (after HI reaction). Unshaded columns denote titres of secretory antibodies $<1:4$ (after HI reaction); the shaded sections denote titres $\geqslant 1:4$. At serum antibody titre <10, $p>0.05$; at titre 10, $p<0.05$; at titre $\geqslant 20$, $p<0.01$.

after the epidemic caused by influenza A(H3N2) virus. This preparation was then introduced into the respiratory tract of mice. Twenty-four hours later they were infected with the A(H3N2) virus. The passive administration of secretory antibodies reliably protected the animals from ten times the LD_{50} of the virus. The fraction isolated from colostrum free of virus-specific antibodies exhibited no such protective activity.

The data cited do not negate the involvement of serum antibodies and cell-mediated immunity mechanisms in protecting the respiratory system from influenza infection. Since mucosal permeability in the lower respiratory tract is much higher than in the upper, isotype G antibodies may be important in preventing diseases of the lungs and bronchi. In the case of inflammatory lesions mucosal permeability increases. This may explain the positive effect of immunization against influenza, through parental administration of an inactivated vaccine, on the severity and incidence of influenza after effects. To sum up, secretory antibodies represent one of the major components in the complex of the organism's specific protective mechanisms for warding off influenza infection.

3. ANTIBODIES IN SERUM

Antibodies to hemagglutinin (HA), neuraminidase (NA), matrix M and NP proteins were found in convalescents. No data on the protective effects of antibodies toward M and NP proteins have so far been obtained. Reverse correlations between the level of anti-HA antibodies and disease incidence and severity have been confirmed in hundreds of cases. The determination of the anti-HA level involves a simple laboratory method used in influenza diagnostics, determination of the immunological efficacy of vaccines, studies into the level of immunity in the population as a whole and the duration of anti-influenza immunity, etc. Immunoglobulin preparations of specific activity, made from sera from immunized donors, are used in the Soviet Union for parenteral administration to children for treating severe forms of influenza. Tissue culture experiments, and the simulation of experimental infection of mice showed that anti-NA antibodies, while not preventing the development of infection, do restrict its spread. It was expected, therefore, that administration of a vaccine which is prepared from a recombinant which contains human influenza virus NA and animal influenza virus HA would produce infection, in immunized people during epidemics, without any clinical manifestations and an immunity which would not be inferior to natural immunity. These expectations, however, did not come true. However, investigations conducted at the All-Union Research Institute of Influenza revealed that persons whose serum contained either moderate or high concentrations of anti-NA antibodies, acquired as a result of illness, but with low titres of antibodies to HA, became infected with influenza A(H1N1) and A(H3N2) viruses less frequently (by 37.5 and 73%, respectively) than people with low titres of both antibody types. In the course of the same observation cooperation between anti-NA and anti-HA antibodies was established. People, whose blood serum contained either moderate or high concentrations of both types of antibodies, became infected with the aforementioned viruses almost 81 and 84% less frequently (Shvartsman *et al.*, 1984). Data on the significant role played by anti-NA antibodies in warding off an infection should be taken into account when selecting strains for vaccine preparation and conducting laboratory assessment of the efficacy of these preparations. Several mechanisms involving anti-HA and anti-NA antibodies seem to play a part in the organism's protection. Antibodies of both types may interact with epitopes of corresponding antigens, may aggregate virions and may prevent their attachment to receptors of the cellular surface membrane.

The first hemagglutinin chain contains four non-overlapping regions, and a neuraminidase molecule contains three regions. Each region contains a group of determinants, i.e. epitopes. The antibodies capable of neutralizing the epitopes of each of the four hemagglutinin regions have been identified. The strain-specific antibodies interact with virions much more actively than do subtype-specific ones (Michell, 1985). However, under certain conditions, the complexes containing influenza A virus and isotype G antibodies, including monoclonals, are capable of attaching themselves to the surface membrane of fibroblasts and penetrating the cytoplasm. In such complexes, however, the virion transcriptase activity was reduced by almost one order of magnitude. The influenza virus envelope contains lipids. This makes possible the activation of complement and destruction of the virions by a complement-antibodies complex. In the course of influenza virus reproduction, its glycoproteins associate with the surface membrane of infected cells. Antibodies to glycoproteins, jointly with complement, may lyse infected cells. Many cells of the lymphoid–macrophage system possess receptors for the Fc fragment of isotype G antibodies. Thus, different immunological reactions may develop as a result of interactions between the antibody molecules and these receptors. One of these reactions, antibody-dependent cellular cytotoxicity (ADCC), will be discussed in the next chapter. The attachment of antibodies to the Fc receptors of phagocytic cells is followed by an increase of their absorptive and metabolizing activity.

Influenza viruses belong to the class of thymus-dependent antigens. In our laboratory, it was found, using a cell culture *in vitro* model, that the least optimal ratio of T to B lymphocytes was two to one. Lymphocytes of both types retain immunological memory. The function of T-helper lymphocytes may be replaced by extracts from the syngeneic and xenogeneic thymus and by lipopolysaccharide as the polyclonal activator of B lymphocytes. After the very first immunization of people with a commercial preparation of inactivated vaccine, the number of immunoregulating suppressor lymphocytes in the peripheral blood increased. This number grew significantly after the second inoculation one month later. Macrophages are important for producing antibodies to influenza virus. They absorb and metabolize both inactivated and living virions, as well as exposing the antigens to T and B lymphocytes. Recently published data underscore the importance of M proteins in generating T-helpers for synthesizing anti-hemagglutinating antibodies. This phenomenon might have provided protection from a lethal influenza infection to mice which had been parenterally immunized with M protein, bound covalently with an adjuvant polyelectrolyte (Petrov *et al.*, 1984).

In humans, antibodies to hemagglutinin could be detected from days 1–7 after infection. By the end of the second or third week their concentration reaches a maximum. It would seem that the use of more sensitive antibody assaying methods would reduce the first of these periods. In any case, an increase in the number of cells forming rosettes, with sheep erythrocytes conjugated with the virus in a lymphocyte suspension isolated from the mediastinal nodes of infected mice, was recorded in the first 12 to 18 hours. This gives grounds for considering antibodies as a factor involved in the organism's protection at early stages of infection. Upon reaching maximum concentration, the number of serum antibodies gradually diminishes. However, they can still be detected over months and even years.

The formation of anti-neuraminidase antibodies proceeds much less intensively than that of anti-hemagglutinins, especially in cases of first infection. The accumulation of both types of antibodies is to a certain extent autonomous. This conclusion was based on the examination of 5500 sera, collected from donors in different seasons and different climatic zones of the Soviet Union. Some 21 to 40% of serum samples contained protective titres (≥1:40) of anti-hemagglutinin, but these samples lacked detectable levels of anti-neuraminidase antibodies. In contrast, high concentrations of anti-neuraminidase antibodies and minimal amounts of anti-hemagglutinin were found in 23% of sera (Shvartsman *et al.*, 1984). This once again underscores that, when studying humoral immunity, antibody determination should not be confined to anti-hemagglutinin antibodies alone.

It has been repeatedly shown that such parameters of antibody genesis as the rate of antibody accumulation, antibody concentration, and the activity spectrum depend on antigen dose and the frequency of immunizations. Guinea pigs were parenterally immunized four times over one-month intervals with inactivated vaccine for which A/Hong Kong/I/68(H3N2) virus was used. The serum collected 3 weeks after the first immunization contained strain-specific anti-hemagglutinins, and after the fourth immunization contained antibodies which reacted with hemagglutinins of all H3 subtype viruses. Following the use of "trivial" methods of antibody isotype determination, it was believed that isotype M antibodies were synthesized during the first infection, those of isotype G developing at the end of the second week and, in the case of reinfection with the same subtype of virus, the formation of isotype G antibodies begins much earlier. After using the immune-enzyme technique, this view was significantly altered. The blood serum of volunteers, inoculated with a live vaccine containing both A(H1N1) and A(H3N2) viruses, showed antibodies of isotypes G, M and A simultaneously. By the end of the second week the concentration of isotypes M and A antibodies declined, while the concentration of isotype G antibodies reached its peak by the fourth week (Couch & Kasel, 1983). Memory of anti-hemagglutinin formation is demonstrated by faster and more intensive production of this type of antibody in cases of reinfection and reimmunization. It would seem that a single infection is sufficient for the formation of an immunological memory. In 1979–1980 young people who in 1977–1978 had suffered from influenza caused by A(H1N1) virus, were observed to produce antibodies in response to the administration of a vaccine prepared from the same virus more actively than their age-mates who had no history of influenza in 1977–1978. The immunological memory is retained for a long period of time. During the mass inoculation of US citizens with a vaccine prepared from A(HswN1) virus, it was elderly and old people who produced antibodies much more actively; they could have been infected with this virus back in 1918–1929, that is, almost 40 years before the inoculations were started. In the literature on the immunology of influenza one may come across the term "original antigenic sin". This term is used to describe intensive production of antibodies to that influenza virus with which a person was infected for the first time ever, or in the case of reinfection with influenza A virus possessing other antigenic characteristics. Studies, conducted at the All-Union Research Institute of Influenza, defined more precisely the notion of "original antigenic sin". Following reinfection, there is increased production of antibodies to all subtypes of influenza viruses to which the given age group of examinees was exposed. The frequency and intensity of increasing anti-hemagglutinin concentrations, to viruses no longer in circulation, are dependent on the degree of antigenic affinity between these viruses and those which are the etiological agents of reinfection. Thus, while examining the serum of people in the older age group, collected prior to 1977, an increase in antibody concentrations to the first-type hemagglutinin was reliably observed less frequently than to the second type during the A(H3N2) virus epidemics. In the course of the same studies, data on the presence of an immunological memory of synthesizing anti-neuraminidase antibodies were obtained. Thus, "original antigenic sin" represents one of the manifestations of immunological memory. Cases of higher titres of antibodies to type one and two of hemagglutinin, on infection with virus containing the HA of type 3, could be taken as proof of common antigen structures in the first chain of the hemagglutinin molecule of all influenza A viruses.

4. CELL-MEDIATED IMMUNITY

Within the limited scope of this section, data on the functions and mechanisms of cell-mediated immunity and its participation in protecting against influenza will be very briefly presented. The basic function of anti-influenza cellular immunity lies in restricting virus reproduction *in vivo*. It is effected by three mechanisms:

1. absorption of viruses and debris of infected cells, this restricts virus spread in the organism;
2. destruction of infected cells before virions are assembled in them;
3. synthesis of biologically active substances of non-antibody nature which activate the virus-absorptive and metabolizing activity of cells by mobilizing them into a focus of inflammation and preventing virus reproduction in cells.

The basic effectors of the first mechanism are macrophages. They initiate the inductive phase of cellular immunity by absorbing viruses, metabolizing them and presenting the antigens to T lymphocytes (Kosyakov *et al.*, 1983). The macrophages' activity increases under the influence of immunity mediators, synthesized by lymphocytes. Macrophages also perform regulatory functions by increasing or suppressing proliferation, differentiation or activity of lymphocytes, sensitized by the antigen. The surface membrane of macrophages contains receptors for Fc fragments of IgG. Therefore, macrophages are mentioned among ADCC effectors. Machrophages, whose activity is enhanced by adjuvants, acquire a capacity for destroying the cells infected with influenza virus, independently, without antibodies or lymphocytes playing any part whatsoever.

Three mechanisms for destroying infected cells, whose effectors are lymphocytes, have been described: cytolysis by cytotoxic T lymphocytes; natural killers (NK); and destruction of cells by lymphocytes jointly with antibodies, or ADCC. Among the major cytotoxic characteristics of T lymphocytes are the identifying of the effector and target cells with respect to glycoprotein I of the main histocompatibility complex (genetic restriction) and the availability of two cytotoxic T lymphocyte types, strain-specific and cross-reactive. These latter two lyse the cells infected with all variants of influenza A virus, but not B virus, and constitute the majority of cytotoxic T lymphocytes in humans (Donerty, 1985).

The mechanism by which cross-reactive cytotoxic T lymphocytes

(CTLs) are formed has been insufficiently studied. Scores of papers published on the subject, including those by Soviet authors, contain evidence that hemagglutinin, M and NP proteins may be involved as antigens in the generation of such CTLs (Andzhaparidze *et al.*, 1983). More and more facts have recently been accumulated on key role of the NP proteins in the formation of cross-reactive CTLs. What is important here, is that interactions between CTLs and infected cells proceed simultaneously with the CTLs producing a number of immunity mediators, including immune interferon. The respective roles of the cytolytic effect and the synthesis of immune interferon in anti-influenza protection have not yet been sufficiently defined. After the adaptive transfer of a clone of strain-specific CTLs, mice were not protected from infection by the heterotypical influenza A virus. This may be taken as evidence of the key importance of cytolysis in the mechanism of action of CTLs. At the same time the adaptive transfer of CTL clones which did not synthesize interferon, was not accompanied by a decline in the intensity of virus reproduction in the lungs of the recipient mice (Donerty, 1985). It is quite possible that immune interferon activates antigens of the H2 complex of epithelial cells, which is a critical event for CTLs affecting the infected cell. Interferon synthesis begins within as little as 2 hours. If infection of humans also triggers an intensive production of immune interferon, this could be taken as an important mechanism of cellular immunity involved in anti-influenza protection. The activation of CTL memory takes place much later, after 4–7 days, when the clinical manifestations of influenza start developing.

Natural killers (NK) are a special group of large granular lymphocytes which do not contain markers of T or B lymphocytes, or macrophages in their surface membranes. The major peculiarities of natural killers are their presence in the organism before infection and their lack of genetic restriction. The capacity of NK for destroying infected cells increases under the influence of interferons. On the other hand, NK are the main interferon producers in humans. Several researchers note a correlation between the decline in the level of natural cytotoxicity and susceptibility to influenza infection. Activation of NK is observed within the first two days of infection. Therefore, one may assume that NK, as well as activated macrophages, represent important components of the organism's early response to virus infection. It would seem that the NK's protective effect is especially important before protective concentrations of antibodies accumulate and CTLs are formed. It is interesting to note the phenomenon of local activation of the NK function. In the lungs of mice infected with influenza virus, such NK activation may be observed from day 2 after infection. However, no intensification of natural killer activity of spleen lymphocytes was detected over the same period. By contrast, if the virus was injected into mice intravenously, NK activity in the spleen was stimulated. Opinions still differ, however, on NK involvement in the organism's anti-influenza protection. It is reported, for instance, that the most marked NK activation is observed in thymusless mice after infection with influenza virus, while these animals continue to release the virus for a long time after infection. In volunteers, infected with influenza A(H3N2) virus, marked clinical manifestations of the disease correlated with a growth in NK activity and interferon synthesis.

A proportion of natural killer cells possess receptors for the Fc fragment of isotype G antibodies. They may be involved in the organism's protection as ADCC effectors. ADCC represents an intermediate link between cellular and humoral immunity. ADCC effectors, primarily lymphocytes, possess receptors for the Fc fragment of isotype G antibodies. The performance of ADCC is not genetically restricted. The specificity of ADCC is determined by antibodies whose F(ab)2 fragment interacts with the virus antigens present on the surface membrane of the infected cell. Complement is not needed for ADCC, and three orders of magnitude fewer antibodies are required than for the lysis of cells by antibodies and complement. This ensures the possibility of ADCC involvement in the organism's protection during the early stages of the infection, as well as the involvement of this reaction in the formation of species-specific immunity. It would seem that ADCC manifestations depend on the properties of both lymphocytes and antibodies. In ADCC, lymphocytes of newborn individuals proved less active than those of adults. Compared to those of young people, the antibodies of older people lysed, acting jointly with lymphocytes, target cells infected with A(H1N1) virus more effectively. On the basis of these data it was suggested that ADCC was involved in the formation in older people of higher resistance to the disease during the 1977–1978 pandemic. The organism's immune response also includes the generation of T lymphocytes as delayed-type hypersensitivity (DTH) effectors. In the case of influenza infection, the main function of DTH lies in developing an inflammatory reaction and concentrating the cellular elements in the focus of inflammation. Inflammation is an essential and important element of the organism's protective reaction. However, when an inflammation becomes excessively aggravated, it may adversely affect the course of the disease. This may explain the delayed development of lung lesions in mice infected with influenza virus and given preparations suppressing the development of immune reactions. To describe the involvement of different cellular immunity mechanisms in the organism's protection, different criteria may be used, such as the correlation between the number of cells of a given type and the manifestations of infection, as well as the effect of the adoptive transfer of cells on its course. Observation of volunteers with low antibody titres established a reverse correlation between the number of memory CTLs and the intensity of virus release after artificial infection. Children with a history of leukosis, who were given cytotoxic preparations, released influenza virus for a long time after catching the disease. However, one cannot exclude that the infection of volunteers and the administration of cytostatic drugs may have affected humoral immunity as well. More convincing results were obtained in experiments on thymusless mice. The roles of NK and ADCC in anti-influenza protection and recovery, have been studied to a much lesser degree. The adoptive transfer of cloned CTLs dramatically rids the organism of the infection provided the infecting dose is not too high. In summing up, one may say that the great importance of cell-mediated immunity lies in its contribution to recovery. However, certain cell-mediated immunity mechanisms, such as synthesis of immune interferon by CTLs, and ADCC, are apparently involved in anti-influenza protection as well.

5. SOME ELEMENTS OF THE INFLUENZA IMMUNITY THEORY AND THE PROBLEM OF A UNIVERSAL VACCINE

Discussion of the elements of the influenza immunity theory should best begin with the issue of the human organism's sensitivity to influenza infection. In isolated human settlements, where influenza cases had not been recorded for dozens of years, introduction of influenza resulted in almost the entire population being severely affected, with the number of lethal cases ranging from 1 to 20%. However, when in 1957 an influenza pandemic was caused by a virus which radically differed from its precursors in the characteristics of both its surface antigens, not more than one-third of the world's population contracted the disease. Again in 1977–1978, during the pandemic caused by the A(H1N1) virus, which was absolutely new to young people born after 1956, the incidence among them did not exceed 15–20%. During both these pandemics, people who fell victim to the disease had it mostly in a mild form, mortality not exceeding hundredth fractions of a percent. These data testify that humans have an inborn high susceptibility to influenza virus, and that people exposed to a wide array of various contacts, enjoy a "universal" immunity to all variants of influenza A virus acquired as a result of reinfections. However, the essence of such a "universal" immunity has not yet been studied. In all probability, memory CTLs, the memory of synthesizing secretory and serum antibodies, as well as the ADCC process may all contribute to it. Man's susceptibility seems to be restricted to the already known pandemic variants of the influenza A virus. This has been proved by "serological archeology" data, i.e. the discovery in elderly people's sera, collected prior to the 1957 and 1968 pandemics, of antibodies to the pathogens of these pandemics, and by the epidemic recirculation of the influenza A(H1N1) virus in 1977. The majority of leading experts in the field share a conviction about the human organism's susceptibility to a limited set of influenza A viruses. The problem of anti-influenza immunity has been a controversial issue. According to some researchers resistance to the disease is retained for 2–7 years if there are no modifying shifts in the pathogen. In 1977, people who were born after 1956 showed an incidence five to eight times higher than elderly people. Studies conducted at the All-Union Institute of Influenza Research showed that reinfection with virus of the same subtype is essential for maintaining immunity for a long period of time. A major feature of immunity to influenza lies in the comprehensive nature of this phenomenon and its reliability. The reliability theory presupposes the reservation of elements which perform similar functions as the major condition for the sustained functioning of complex systems. Secretory and serum antibodies, serum antibodies and interferon, CTLs and macrophages, all interact in a pronounced way in protecting against influenza infection. The biological activity of a virus, localized extracellularly, is neutralized by antibodies in respiratory tract secretions and antibodies in blood serum, while the destruction of infected cells is effected by antibodies jointly with complement and lymphocytes, CTLs, NK and activated macrophages. The principal limitation of modern anti-influenza vaccines is the excessively strict specificity of their protective effect. No wonder, therefore, that researchers in different countries strive to devise "universal" vaccines whose effects would not depend on the variability of the antigen characteristics of viruses. Two major trends in the research in this field — "virological" and "immunological" — should be singled out.

The first, represented in the Soviet Union by the work of Petrov, the late Victor Zhdanov and their co-workers (Petrov *et al.*, 1984), is mainly oriented towards the use of viral structures, common to all variants of influenza A virus, for building up the material base of universal vaccines, such as M protein, conservative sections of the first hemagglutinin chain, and the polypeptide section of fusion of the second hemagglutinin chain. To enhance their immunogenecity these structures are bound up with adjuvants or, as in the Soviet Union, with synthetic copolymers of polyelectrolytes. These drugs were parenterally administered to mice. In two weeks the animals acquired resistance to different variants of influenza A virus.

The second trend is being developed in our laboratory. It is based on the notions of the existence of natural "universal" immunity, the comprehensive nature of non-susceptibility and cooperative interaction between the components in this complex, the presence of common structures in the first hemagglutinin chain of all influenza A viruses, and the participation of hemagglutinin, neuraminidase and M and NP proteins in immunogenesis. The preparation for influenza prophylaxis,

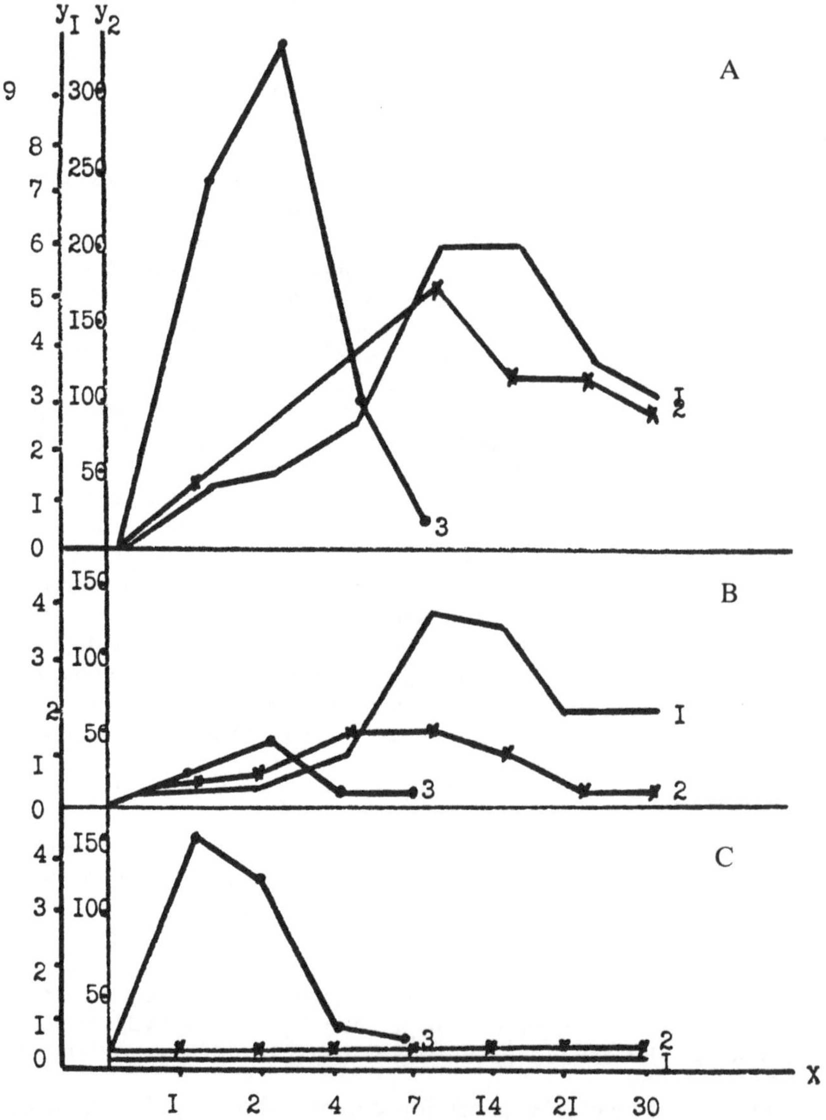

Figure 3 Synthesis of serum and secretory antibodies and interferon in mice after the intranasal administration of KPG and its components: A (top), for mice given KPG; B (middle), given IIV; and C (bottom), given IIIS. *Ordinate* y_1, geometric mean antibody titres (in log 2); *ordinate* y_2, geometric mean interferon concentration in lung tissue homogenates (in IU); 1, titres of blood serum antibodies; 2, titres of secretory antibodies; 3, interferon concentration.

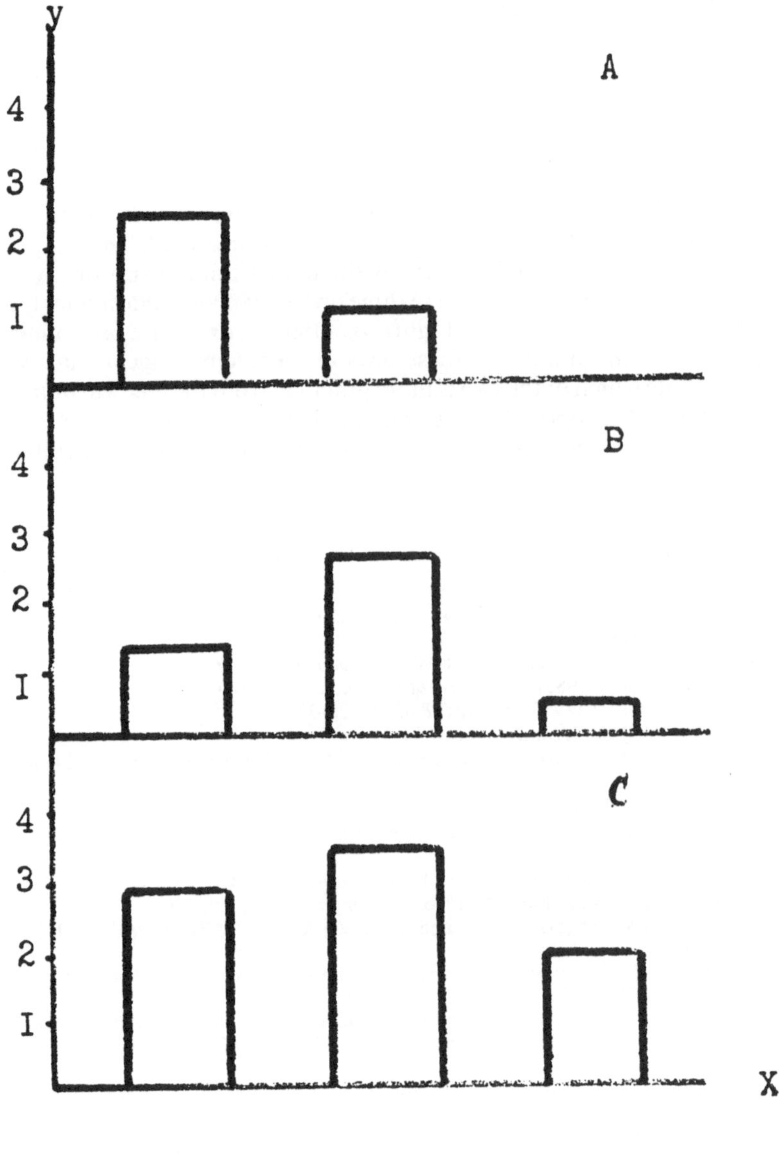

Figure 4 Efficacy of protection in mice infected with virus A/PR/8/34(H1N1) after administration of KPG and IIV, prepared from the A(H3N2) virus, or their components: A (top), mice given IIIS; B (middle), given IIV; C, (bottom) given KPG. *Ordinate*, difference in infection resistance between test and control animals (in log ID_{50}); *absissa*, period of infection after administration of preparations.

known under its tradename as KPG (Shvartsman *et al.*, 1984), consists of two components: inactivated influenza vaccine (IIV) and an interferon inducer or immunogenesis stimulator (IIIS). It was initially supposed that the function of IIV lies in inducing the synthesis of antibodies and interferon, while that of IIIS would be to induce the synthesis of interferon and stimulate the synthesis of antibodies. This supposition was confirmed after intranasal administration of such drugs to mice (Figure 3). In the course of the main test it was found that immunization by KPG resulted in the development of heterotypical resistance after the first 24 hours of the drug's administration. At all stages of observation it was two to three orders of magnitude higher than in the control group animals (Figure 4). After the drugs are pronounced safe for tests in volunteers, these tests and epidemiological trials will demonstrate which of the two is more practical from the medical point of view. It may be hoped that the wide application of universal vaccines would open up the prospect of turning influenza into a controllable infection.

REFERENCES

Andzhaparidze, O.G., Sorokina, Yu.D., Chernushkina, A.V. *et al.* (1983) *Vopr. virusol.*, **2**, 136–139 (In Russian).

Angelova, L. and Shvartsman, Ya.S. (1982) *Immunology*, **46**, 183–188.

Couch, R. and Kasel (1983) *Annu. Rev. Microbiol.*, **37**, 529–549.

Dhar, R. and Ogra, P. (1985) *Br. Med. Bull.*, **41**, 28–33.

Donerty, P. (1985) *Br. Med. Bull.*, **41,** 7–14.

Kosyakov, P.N., Berdinskikh, M.S. and Kisipova, A.S. (1983) *Immunologiya*, **1**, 34–38 (In Russian).

Michell, D. (1985) *Br. Med. Bull.*, **41**, 80–85.

Petrov, R.V., Zhdanov, V.M., Kabonov V.A. *et al.* (1984) *Dokl. Akad. Nauk SSSR*, **277**, 720–722 (In Russian).

Shvartsman, Ya.S., and Grigorieva, S. (1979). *J. Infect. Dis.*, **140**, 837–843.

Shvartsman, Ya.S. and Zykov, M. (1976) *Adv. Immunol.*, **22**, 291–330.

Shvartsman, Ya.S., Grigorieva, S.K. and Maltseva, A.I. (1977) *Zh. Mikrobiol.*, **9**, 53–56 (In Russian).

Shvartsman, Ya.S., Grigorieva, S.K. and Smorodintsev, A.A. (1978) *Dokl. Akad. Nauk SSSR*, **242**, 974–976 (In Russian).

Shvartsman, Ya.S., Naikhin, A.N. and Korchanova, N.L. (1984) *Problems of Influenza Immunology*. Moscow: The All-Union Research Institute of Medical Information (In Russian).

Shvartsman, Ya.S., Ispolatova, A.V., Tsibulskaia, N.V. *et al.* (1982) *Vopr. Virusol.*, **4**, 142–146 (In Russian).

Shvartsman, Ya.S., Shuratov, I.Kh., Naikhin, A.N. *et al.* (1985) *Immunology of Influenza*. Alma Ata: Nauka (In Russian).

Shvartsman, Ya.S., Tsibulskaya, N.V., Ispolatova, A.V. *et al.* (1984). In *Immunity and Non-specific Resistance to Influenza and Other Diseases of the Upper Respiratory Tract*, pp. 5–17. Leningrad: The All-Union Research Institute of Influenza (In Russian).

Tsibulskaya, N.V., Popova, T.L. and Ispolatova, A.V. (1983) *Zh. Mikrobiol.*, **7**, 81–83 (In Russian).

Sov. Med. Rev. E. Virology Reviews, Vol. 3. 1990, pp. 19–35
Photocopying permitted by license only

FORMS AND MECHANISMS OF INFLUENZA A VIRUS CIRCULATION IN THE BODY

A.F. FROLOV

Research Institute of Epidemiology and Infectious Diseases, Kiev

Abstract

This article describes the results of many years of systematic investigation, by the author and coworkers, into influenza A virus circulation within the host body and its population. It has been established that the agent may occur in the body in three principal forms: namely, a free-circulating form, an infectious complex, and an immune complex. The complexes arise when influenza virions combine with specific immunoglobulin molecules and as a result the virus still circulates at later stages of infection (1 to 6 months). A free-circulating form is recognized primarily in the acute phase of influenza (7 to 10 days). Evidence is cited that some components of influenza viruses (M, NS and other proteins) change during their continuing survival in the body and the related role of defective-interfering (DI) virus particles is discussed.

1. INTRODUCTION

Studies into the circulation of influenza virus in acute influenza infection (*in vivo*) using conventional virological techniques have determined the timing of the release of the virus from the body and its main properties. In the past there had been occasional reports that the duration of influenza virus release from the body may exceed the generally acknowledged period of 7 to 10 days. More recently, while studying the co-carcinogenic action of influenza virus, we have hazarded a conjecture to the effect that "the tumorigenic characteristic of influenza virus is most marked in cases where the productive type of the infectious process gives way to the abortive one, and the course of the disease is prolonged. In such cases, freeing the body from influenza virus is retarded so that the latter will persist in a viable cell for a longer time." (Ritova, 1969). Something like this may well occur following preventive administration of live attenuated virus (Frolov, 1973).

Research into the distinctive circulation of influenza A virus within the body at different stages of the infectious process in terms of pathogenetic mechanisms whose operation involves humoral immunity factors, prompted us to propose the possible formation of antigen–antibody complexes capable of circulating in the body for a long time (Frolov *et al.*, 1986). As with some other microorganisms, influenza virus forms two types of complexes: the infectious, arising when the antigen is in excess while the antibody is in short supply; and the immune, occurring when the ratios between these components become equal or change in favour of antibodies (specific immunoglobulins). The first type of complex is identified largely in the acute phase of influenza infection and this complex retains the infectious and hemagglutinating properties of the virus because the virion surface structures responsible for these properties are incompletely blocked by immunoglobulins. The second type of complex lacks these properties, since the virion's activity is practically completely blocked by antibodies. These structures, in our opinion, are not rigidly stable and are capable, under certain conditions, of interconversion. The identification of infectious and immune complexes depends, however, on those differences in properties. With this in mind, we employed an antiglobulin serum whose action on the infectious complex deprived it of infectious and hemagglutination properties, whereas exposure of the immune complex to this serum resulted in the appearance of these properties – apparently due to the binding of some immunoglobulin molecules on the surface of the virion and the freeing of its surface active centres. In our strategy, 0.3 mL of antiglobulin serum is added to 0.3 mL of the test material. After incubation for 45–60 min at 37°C, the mixture was centrifuged for 5 min at 2000 rpm. 0.1 mL of the resulting supernatant was analysed by the hemagglutination (H) reaction, and 0.2 mL of the residue is inoculated into the allantoic cavity of 9 to 10-day-old chick embryos which are then placed in an incubator for 72 h at 33°C. The isolated viruses are identified by the hemagglutination-inhibition reaction, complement fixation, hemadsorption inhibition on a cellular monolayer and passive hemagglutination.

2. EXPERIMENTAL WORK IN ANIMALS ON FORMS IN WHICH THE VIRUS CIRCULATES

We have studied various forms of influenza virus circulation in experiments with CBA and BALB/c mice (Frolov and Garrilov, 1981). The animals were infected with the A/Hong Kong/1/68 (H3N2) agent recovered from a patient in accordance with WHO criteria and adapted to mice. The influenza virus infectivity titre was 10^3 log EID_{50} and the hemagglutination titre was 512 HAU mL^{-1}; the animals were given the virus intranasally under ether anesthesia in doses from 0.05 to 0.1 mL. The mice were sacrificed on the 1st, 3rd, 5th, 7th, 9th, 12th, 15th, 18th and 21st days after infection. Lung, brain, spinal cord, liver, kidney and spleen tissue, as well as blood serum, were used for examination under sterile conditions. Upon careful washing, grinding and addition of isotonic sodium chloride solution, organ tissues yielded a 10% suspension which was centrifuged for 10–15 min at 1500 rpm. The supernatant was aspirated and subjected to examination. The samples were each divided into two parts; one group of biosubstrates was treated with antiglobulin serum and the other was untreated. The results obtained are presented in Table 1.

It can be seen from Table 1 that treatment with antiglobulin serum strongly affected the results of assaying influenza virus infectivity in the test material. The results are best considered by dividing the biosubstrates into three groups. In the first group (serum, lung and liver), the application of antiglobulin serum resulted in a reduction of infectivity as measured by a routine virological method (infecting 9 to 10-day-old chick embryos). Depending on the organ and the observation period, the reduction varied between factors of 1.1 and 2, pointing to the presence in the biosubstrate of the influenza virus infectious complex, neutralised by the antiglobulin serum, and of antibody-unbound virions (arbitrarily designated as a free-circulating form) unaffected by the antiglobulin serum.

The second group (spleen) was characterised by the fact that exposure of the derived biosubstrates to the antiglobulin serum led to the discovery of infectivity which was not detectable by conventional virological methods. This suggested that influenza virus is present in

Table 1 Levels of influenza virus infectivity[a] in the organs of CBA mice at different times after infection with A/Hong Kong/1/68 (H3N2) virus with (+) and without (−) antiglobulin serum treatment of tissue prior to assay.

Group	*Tissue*	*Antiglobulin serum*	*Time after infection (days)* 1	3	5	7	9	12	15
I	Blood serum	−	6.8±0.4*	1.1±0.3*	0	0	0	0	0
		+	3.7±0.4	0	0	0	0	0	0
	Lungs	−	11.3±0.3*	12.8±0.4*	10.3±0.3*	6.2±0.3*	3.8±0.4*	1.2±0.4*	1.0*
		+	7.2±0.6	8.8±0.3	6.5±0.4	4.4±0.6	1.9±0.3	0	0
	Liver	−	2.5±0.4*	8.8±0.3*	3.1±0.4*	2.3±0.4*	1.8±0.4*	0	0
		+	1.3±0.3	4.5±0.4	2.1±0.3	2.1±0.3	0	0	0
II	Spleen	−	0	0	0	0	0	0	0
		+	1.6±0.4	2.1±0.3	0	0	0	0	0
III	Brain	−	1.4±0.3*	0	0	0	0	0	0
		+	0	1.6±0.4	2.8±0.4	1.8±0.4	0	0	0
	Kidneys	−	0	3.9±0.4*	0	0	0	0	0
		+	0	1.9±0.3	1.1	0	0	0	0

[a] Infectivity measured by standard methods as rate of infection in chick embryos: 0 = No infectiousness found in substrate; Values are means ± standard error.

* Infectious complex present.

spleen tissue as an immune complex, the exposure of which to the antiglobulin serum brings about the deblocking of the virion's surface structures responsible for its infectivity, which, though proved to be low (1.6–2.1 log EID_{50}), was sufficient for determination.

The third group (brain, kidney) was distinguished by the fact that at the early stages of examination, within 1–3 days of infection, the influenza virus infectious complex (a reduction in infectivity following treatment of the material with antiglobulin serum) and its free-circulating form (residual infectivity not neutralizable on exposure to antiglobulin serum) were detectable in the animal tissues. But from day 3 (brain) and day 5 (kidneys), influenza virus was found in these organs only in the form of immune complexes.

Thus, the data obtained suggested that in experimental animals influenza virus circulates in three forms: a free-circulating form, an infectious complex, and an immune complex. These represent its combination with specific antibodies. In the acute phase of influenza infection, the agent was detectable in lungs for the longest period of time (for the 15 days prior to termination of the experiment). It should be stressed here that while the infectious complex and free-circulating form were both detected up to 9 days, the infectious complex alone was detected later. Liver showed a similar pattern, while being less pronounced as to time (up to 9 days) and virus activity. The parental influenza virus A/Hong Kong/1/68 (H3N2) used to infect the mice was isolated (during an observation period of 15 days) from all the biosubstrates derived from the animals infected in this experiment.

3. INVESTIGATION OF VIRUS CIRCULATION IN HUMAN SUBJECTS

In addition to animal experiments, we examined 161 humans. Of that number, the diagnosis of influenza was established in 79 individuals on clinical, epidemiological and laboratory evidence; 25 exhibited disorders of the nervous system, presumably of viral etiology; and 57 were convalescents with a past history of influenza.

Group 1: the influenza cases were kept under observation in January–March 1979;
Group 2: those with an impaired nervous system were observed in November 1976–April 1977;
Group 3: convalescents were examined in February–April 1977.

Traditional methods such as passaging in 9 to 10-day-old chick embryos with a view to detecting the agent and estimating its degree of infectiousness, the hemagglutination (H) reaction and the hemagglutination-inhibition (HI) test, as well as the antiglobulin serum treatment (described above), were employed for a virological examination of the materials obtained from the persons included in all the three groups.

3.1 Group 1: Influenza patients

Among patients of the first group, influenza virus classified as A/Texas/77 (H3N2) was isolated virologically in 39.1% of the cases and from nasopharyngeal washings in 13.1%; clinical diagnosis was confirmed by serological evidence in 31.5% of the cases. The results of identifying antigen–antibody complexes in this group of patients are tabulated in Table 2.

Table 2 Blood antigen–antibody complex levels in influenza patients.

Blood component	*No. of samples*	*No. of influenza virus strains isolated*	*Infectious complex*			*Immune complex*		
			Days[a] 1–3	*Days 4–7*	*Days 8–12*	*Days 1–3*	*Days 4–7*	*Days 8–12*
Plasma	79	20	13/16[b]	3/3.8	0	4/5.1	0	0
Leukocytes	79	21	5/6.3	3/3.8	0	4/5.1	6/7.6	3/3.8
Erythrocytes	79	7	0	2/2.5	0	3/6.3	0	0

[a] Days are days of disease.
[b] Numerator shows absolute values of positive findings; denominator is the percentage of the total number of samples investigated; 0 = no complexes found.

As can be seen, influenza virus circulates in patients as both immune and infectious complexes. The latter were detected in the plasma and white blood cells in a large number of subjects as early as the first day of illness. In erythrocytes, the infectious complex was found on the fourth day of illness. The immune complex of influenza virus was detected for the most part in the formed elements of the blood, in leukocytes it persisted for as long as 12 days of the experiment in fairly high titres. Fewer patients showed immune complex in their plasma. In most cases influenza virus–antibody complexes ceased to be detected during the period of observation (21 days); however, in one out of five examinations they persisted for as long as 30 days. Whatever the method used, all influenza virus specimens isolated during this phase of disease were identified as the A/Texas/77 (H3N2) strain that had caused an outbreak during which these patients were examined.

Thus, the agent may occur in the blood of influenza patients as a free-circulating form and as infectious or immune complexes. Depending on the activity of the body's resistance systems and the particular stage of infection, these forms may, in our opinion, be interrelated in a "virus free-circulating form–infectious complex–immune complex" cycle. We believe that in the acute phase of the disease, the agent is found chiefly as a "free-circulating form", permitting its isolation by classical virological methods up to 7, and on occasion 12 days from the onset of illness. As processes associated with the mechanisms of cell-mediated and humoral immunity evolve, the body's specific immunoglobulin levels increase, altering the proportions in the "influenza virus antigen–antibody" complex and giving rise first to infectious and, subsequently to primarily immune complexes. When the infectious process is uncomplicated and cyclic at the convalescence stage, these are the principal forms in which the agent circulates and identifying it necessitates special procedures, notably the use of antiglobulin serum. The complexes are eliminated by the body, deposited in cells of the reticuloendothelial system and renal tubules, or circulate continuously in the bloodstream.

The examination of 75 patients revealed that the infectious complex was most often (in 16 out of 26 patients) found in the blood plasma predominantly early (1st–3rd day) in the course of influenza. In leukocytes it was detectable to a lesser degree, whereas in erythrocytes it was found in only two out of 26 cases at 8–12 days of illness. In 18 out of 27 cases, the immune complex was detected primarily in the leukocyte fraction for a long period of time. These complexes were detectable not only within the period of 8–12 days but also at day 21 and later. A total of 53 influenza virus strains were isolated from the patients, 26 being recovered from the infectious and 27 from the immune complexes. All of them were identified as either A/Texas/77 (H3N2) or A/Khabarovsk/77 (H1N1).

On the basis of the circulating form in which the virus was detected, the subjects were divided into three sub-groups characterized by the presence of:

1. the predominantly free-circulating form (32 persons);
2. the infectious complex (16 persons);
3. the immune complex (27 persons).

Patients of the first sub-group with the agent circulating free in the blood serum were mostly admitted to hospital 1–3 days after the onset of disease; only three out of 32 cases were hospitalized at a later date, during the second week. In this group, the acute onset of influenza occurred in 95% of the subjects without being preceded (except for two cases) within the immediately previous 2–3 months by acute respiratory viral infections (ARVI). In 18.6% of the cases illness was severe. Sixteen out of 32 patients responded by running a temperature of 39°C or higher, with a subfebrile response in two cases only. With uncomplicated influenza, fever lasted 3.4 days, with complicated influenza 4.8 days. Other manifestations of the toxic syndrome, such as dizziness and headache, were observed, on average, over a period of 2–4.2 days in uncomplicated influenza and 2.3–4.3 days in complicated influenza. Complications in this group occurred in 48.4% of the cases and were marked by pharyngitis averaging 7 days in duration, pneumonia (24.24%), sore throat (9.1%), maxillary sinusitis (6.1%) and meningitis (9.1%). In five patients, neurological complications were limited to meningitic symptoms that became apparent when influenzal intoxication was most clearly manifest and disappeared as it diminished. Hospitalization of this group of patients with uncomplicated and cyclic influenza lasted for 6.2 bed days, with complicated influenza 11.6 days, and in cases of nervous system disorders 12.6 bed days.

In the second sub-group of subjects, where influenza virus circulated as an infectious complex, acute onset of the disease was observed in 63% of the cases, with three persons having a past history of ARVI of unspecified etiology. Affected persons (except one) were hospitalized within 2 or 3 days of the early manifestations of influenza. Its severity was for the most part (87.5%) medium and in one case grave. In four out of 16 subjects fever was upwards of 39°C, and in six patients the illness ran a subfebrile course. Uncomplicated influenza lasted as long as 4.3 days, complicated influenza for 5.1 days. Asthenia persisted for 4.4 days, headache 2.2, and bodily rheumatic pain 2 days; in complicated influenza, they lasted 6.1, 4 and 2.1 days, respectively. Pharyngitis and concomitant coughing were observed over a period of 6.6 to 7.6 days. Eleven out of the 16 patients of this group (68.75%) showed influenza complicated by pneumonia (31.25%) and sore throat (37.5%), and one person developed infectious allergic myocarditis. In uncomplicated and cyclic influenza, hospitalization of the patients of the second sub-group averaged 7 bed days, and in complicated acyclic influenza, 10.8 bed days.

In the third sub-group of subjects with influenza virus circulating largely as an immune complex, acute onset of the disease was noted in 72% of the hospitalized patients, and 22.2% of the cases had suffered from ARVI 2 or 3 months prior to this illness. One third of the cases were admitted to hospital in the second week of illness, the rest were admitted earlier. Influenza was severe in 25.9% of the cases, mild in 11.1%, and medium in 62.9%. A temperature of over 39°C was noted in 33.3% of the cases, and subfebrile temperature in two (7.4%) out of 27 cases. Uncomplicated and complicated influenza lasted 3.5 and 4.8 days, respectively. Other manifestations of influenzal intoxication, such as headache, dizziness, and body rheumatic pain were observed over periods of 2.5, 3 and 2.5 days, respectively, in patients with a cyclic infectious process, and over periods of 4.7, 3.2 and 2.6 days in patients with an acyclic infection. In the patients of this sub-group, there was a retarded and torpid abatement of symptoms whose intensity periodically increased. Sixteen (59.26%) out of 27 persons exhibited complications such as pneumonia (29.6%), sore throat (18.5%), maxillary sinusitis (7.4%), and central nervous system disorders noted in 22.22% of the cases. Moreover, the meningeal syndrome (three persons) marked by the symptoms of hypertension and meningitis, severe headache, nausea, vomiting, positive Kernig's and Brudzinski's signs, and occipital muscle rigidity, was observed along with the meningo-encephalitic syndrome and meningo-encephalitis with signs of focal brain injury, such as nystagmus, anisocoria, Barre's signs, Babinski's signs, with abdominal and plantar reflexes diminished or entirely absent. Patients of this group (11.11% of the cases) developed severe cardio-vascular complications in the form of clinically manifested toxic and infectious myocarditis. Persons with the immune complex showed pharyngitic symptoms lasting, on average, for up to 7.7 days.

It should be noted that relapses and a protracted course of the disease were observed in the hospitalized patients of this group alone, two patients displaying recurrent post-influenzal viral–bacterial pneumonias: one had a recurrence of the meningo-encephalitic syndrome, and the other a protracted lobar pneumonia.

In complicated influenza, hospitalization of the third sub-group of patients amounted to 15.5 bed days; in uncomplicated influenza, 7.3 bed days.

As seen from our data, the circulation of influenza virus mostly in a free, antibody-unbound form, is typical to a greater extent (95%) of individuals with an acute onset and course of illness. The temperature exceeded 39°C in half the patients of this sub-group, fluctuating between 37.5 and 38.9°C in 43.8% of the cases. By contrast, an increase of "influenza virus–antibody" complexes as a proportion of the total circulating fraction (warranting the formation of the second and third sub-groups) was attended by a drop in the proportion of patients with acute onset of illness to 73–72%. Accordingly, the number of persons with a subacute process increased to 27–28.9% (5% in the first sub-group). But all this notwithstanding, in cases where influenza virus circulated mainly as an immune complex, the number of severe forms soared to 25.9% (compared to 12.5% in the first sub-group) while the proportion of complications increased only slightly to 59.3–68.7% from the 48.4% in patients with free-circulating virus. What is more, there was a change in the pattern of complications. While the percentage of pneumonias, approximately the same in all sub-groups (24.2–29.6), was substantially unaffected, complications affecting the nervous system, along with clinically pronounced cardiovascular disorders, occurred much more frequently (22.2%) in patients with the continuous type of immune complex circulation. These disorders were also noted in cases where infectious complexes were continuously produced and, in our opinion, may be due to the distinctive action exerted on the body by antigen–antibody complexes deposited in the cells of the reticulo-endothelial system, Kupffer's cells, and vascular walls. Recurrent influenzal complications were noted in the third sub-group alone (11.1%), and hospital treatment of these patients, especially with complications, proved to be the longest – 15.6 bed days as against 11.6 bed days in the first sub-group.

The formation of antigen–antibody complexes is a common biological feature representing, with a cyclic infectious process, a defence mechanism used by the body to combat the infecting agent. It would be expected that the agent combined with specific antibodies (immunoglobulins) and circulating as immune complexes, would be eliminated from the blood-stream and the body. However, our results show that this is not always so. In these cases, the continuous circulation of "influenza virus–antibody" complexes may be taken as an unfavourable prognostic sign indicative of disturbances in the body's antiviral resistance system. Whether or not the antigen–antibody complex continues to circulate, and what effects this may have on the patient, will depend on the particular patient's immunogenetic characteristics and the properties of the infecting agent.

3.2 Group 2: Patients with neurological disorders

Examination of the cerebrospinal fluid in 25 patients with disorders of the nervous system revealed influenza virus in eight: serotype A0/PR/8/34 (H0N1) in five cases, A1 in two and A2 in one (Table 3). The agent occurred in the fluid only as an immune complex.

In individual cases, virus was identified in serum even 120–150 days after the onset of the disease.

Virological analysis of the cerebrospinal fluid of our original Group 2 patients (disorders of the nervous system) resulted in isolation of one case of A/Kiev/168/77 (H0N1) influenza virus being isolated in a meningo-encephalitic patient on the 150th day of illness.

3.3 Group 3: Convalescents

The examination, in February–April 1977, of convalescents with respect to factors capable of influencing the body's immunological status revealed some of the features of influenza virus circulation in this group.

As seen in Table 4, an examination of 57 convalescents at different

Table 3 Virological analysis of cerebrospinal fluid from 25 patients with disorders of nervous system.

Form in which virus found	*Influenza virus isolated*					
	Total	*A0*	*A1*	*A2*	$HA.mL^{-1}$	$log\ EID_{50}$
Free-circulating form	—	—	—	—	0	0
Infectious complex	—	—	—	—	0	0
Immune complex	8	5	2	1	28.0 ± 0.35	3.96 ± 0.54

times after the illness, culminated in 25 cases in which A0, A2 and B influenza viruses were isolated. Their isolation from the human body after as long as 144 days is interesting in itself and points to the possibility of continuous influenza virus survival in the human body. All the examined virus carriers had frequently suffered from ARVI and "the flu", which cannot but be the result of the continuous circulation of the agent. Influenza viruses of A0 and A2 serotypes were detected in the body only as immune complexes and were not generally identified until repassaging on chick embryos. Their hemagglutinin titres ranged between 2 and 16 HAU mL^{-1}, and infectivity titres between 2.5 and 3.5 ± 0.4 log EID_{50}; in contrast, the type B agent was found as immune and infectious complexes as well as the free-circulating form. Its hemagglutinin titre was in the range of 2 to 14 HAU mL^{-1}, and infectivity varied from 1 to 3 ± 0.3 log EID_{50}.

Table 4 Results of virological analysis of blood samples from 25 out of 57 convalescents with a past history of influenza.

Circulating form of virus	*No. of strains isolated*	*Days elapsed since illness*			
		15–30	*35–50*	*55–70*	*95–144*
Free-circulating	2	1/21[a] (1BHK)	0/10	1/14 (1BHK)	0/12
Infectious complex	6	1/21 (1BHK)	0/10	3/14 (3BHK)	2/12 (2BHK)
Immune complex	17	5/21 (1BHK 2AO 2ATEX)	3/10 (1AO 2BHK)	4/14 (1AO 3BHK)	5/12 (5BHK)

[a] Numerator represents the number of positive findings; denominator the total number of subjects; the number and type of influenza strains are in parentheses.
BHK = B/Hong Kong strain; A0 = AO/PR/8/34(H0N1); ATEX = A/Texas (H3N2).

4. FURTHER INVESTIGATIONS

In addition to establishing that influenza virus continues to circulate as antigen–antibody complexes in convalescents and in patients with meningo-encephalitis of influenzal etiology, we conducted a series of investigations during the influenza epidemic of 1976–1977, known to be caused by type A2 and B agents. During these investigations the strain A/PR/8/34 (H0N1) was isolated. This circumstance we believe to result from the persistence of the agent in the body, possibly circulating in the form of antigen–antibody complexes, which may give rise to changes in the virion's structure. These processes are far more complicated than is currently realized. However, it stands to reason that during the period preceding the appearance of A/USSR/80/77 (H1N1), the influenza virus population underwent dramatic changes involving the emergence, by no means accidental, of A/PR/8/34 (H0N1). The failure of A(H0N1) to emerge among the "dominant" strains in lieu of or similar to A (H1N1) is yet to be explained.

Following the 1976–1977 influenza epidemic, caused primarily by the type B agent, we examined, at different dates, 20 patients with chronic non-specific pulmonary illnesses (Frolov *et al.*, 1978). We analysed blood serum samples treated as described above. A total of four influenza virus strains were isolated whose main characteristics are listed in Table 5. It turned out that influenza virus circulated in the subjects as an infectious complex and was detectable only using antiglobulin serum; antigenically, it was identified with the A/PR/8/34(H0N1) type.

The finding that the influenza virus was a component of antigen–antibody complexes prompted us to look at the relationship between their circulation and the course of the infectious process (Frolov *et al.*, 1982). Continuous influenza virus circulation in the body proved to affect some of the properties of these complexes. This conclusion was drawn from studies on C57B1 mice infected with the adapted A/Hong Kong/1/68 (H3N2) influenza virus obtained from the Ivanovsky Institute of Virology, USSR Academy of Medical Sciences. The animals were infected with a standard virus (SV) and the agent's population was enriched with defective–interfering (DI) particles responsible for continuous virus circulation in the test animals. The SV infectivity titre was 8.5 log EID_{50} per 0.1 mL; the hemagglutination titre was 512 HAU; and the LD_{50} 4.5 log. With DI particles, the corresponding figures were 4 log EID_{50} per 0.1 mL and 256 HAU; the ID/HAU ratio being 5×10^2. Each mouse was intranasally given 0.05 mL of virus-containing material at a titre of 4 log EID_{50}. Control animals received 0.05 mL of an isotonic sodium chloride solution as an alternative to virus. The animals were sacrificed on the 3rd, 5th, 7th, 15th and 20th days and so on up to day 60, and then monthly. Lungs, liver, kidneys and spleen were extracted in

Table 5 Biological and antigenic characteristics of influenza virus strains isolated from 20 patients with chronic non-specific pulmonary illnesses following the 1976–1977 influenza epidemic

Biological properties		*Antigenic properties*				
Reproduction			*Titre in HI Test*		*Titre in NI (neuramenidase inhibition) test*	
in chick embryos (log EID_{5-})	*Neuraminidase activity (enzyme U mL^{-1})*	*Reference sera virus type*	*With homologous virus*	*With test variants*	*With homologous virus*	*With test variants*
2–8	0.046–0.050	A/PR/8/34 (H0N1)	640	320–1280	1:76	1:11–14
		A/Sca/53 (H1N1)	640	0	—	—
		A/Ned/56 (H1N1)	640	0	1:54	0
		A/Singapore/57 (H2N2)	640	0	1:320	0
		A/England/12/64 (H3N2)	640	0	—	—
		A/Hong Kong/1/68 (H3N2)	640	0	1:815	0
		A/England/42/72 (H3N2)	320	0	—	—
		A/Victoria/35/72 (H3N2)	1280	0	—	—
		A/Port Chalmers/1/73 (H3N2)	2560	0	1:830	0
		A/Victoria/3/75 (H3N2)	2560	0	—	—
		X-97 (H0N2)	2560	160–2560	1:1000	0
		X-2967 (H0N2)	5120	1280–10 240	1:3200	0
		B/USSR/0443/69	640	0	—	—
		B/Hong Kong/5/72	640	0	—	—
		B/Japan/73	1280	0	—	—
		B/Hong Kong/7/75				
		Sendai virus	640	0	—	—
		Newcastle disease virus	1280	0	—	—

sterile fashion, and their tissues were tested for virus or its components. A 10% organ suspension was used in infecting chick embryos, and in hemagglutination and CF (compliment fixation) tests. The isolated strains were identified by matching against standard diagnostic and rat sera as well as a serum to Sendai virus. Also determined were neuraminidase activity, its thermosensitivity and that of hemagglutinin. Details of the methods used in investigating the synthesis of virus-specific proteins and analysing virion proteins are set out elsewhere (Frolov *et al.*, 1981).

Infecting the animals with a standard influenza virus killed them in 72% of the cases, and exposure to virus enriched with DI particles killed 52% within the same observation period. Interestingly, mice infected with DI virus produced the agent continuously (for up to 6 months).

In the acute phase of infection (at 7–8 days) viruses identified as A/Hong Kong/1/68 (H3N2) were isolated by routine virological techniques. Isolation of virus at later dates called for additional methods and procedures, included washing organs free of blood serum antibodies, exposure of tissue to ultrasound (UZDN-1 unit, 35 kHz, 90 s), and then to an equal volume of trypsin (5 $\mu g\ mL^{-1}$). The agent was most commonly found in the spleen. In all, 12 strains were isolated, of which eight were fixed in passages on chick embryos and formed two groups.

The first group incorporated five strains obtained 2 to 3 months after the mice were infected with DI-enriched virus; in their principal properties – infectivity, hemagglutinin and neuraminidase activity – these strains differed only slightly from the parental strain. The second group included three strains isolated 5 to 7 months after beginning the experiment. They were distinguished by their low pathogenicity for mice (Table 6) and higher hemagglutinin thermal stability; upon heating at 56°C for 30 min, the figures were 6.5 and 3.12, respectively, compared to the parental virus. The antigenic structure of these viruses was identified as H3N2.

Table 6 Lethal effects in C57B1 mice of exposure to influenza virus and its persisting variants[a].

Virus	*Dose (log EID_{40})*	*Death rate %*	*Dose (log EID_{50})*	*Death rate (%)*
Original A/Hong Kong/1/68	4.0	51.52±5.7	8.5	81.0±4.4
PS variant (2-month infection)	4.0	7.5±2.8	9.5	16.3±3.7
VS variant (6-month infection)	4.0	0	9.5	3.75±0.39

[a] Mice were given 0.1 mL of virus-containing material

Studies into the electrophoretic mobility of virus-specific proteins showed that in a strain isolated 2 months after infection there was no difference from the parental virus. In contrast, the virus isolated after 6 months did show differences in terms of the mobility of *M* and *NS* proteins: there was a rise in the mobility of the *M* protein and a drop in the mobility of the *NS* (Figure 1).

We consider these changes to be the result of substantial deviations in the genes responsible for the synthesis of the structures mentioned. This may be the cause of phenotypic changes in the properties of viruses isolated in the later stages of influenza infection. However, the established changes may also be due to the impaired activity of other genes, an impairment not reflected in the electrophoretic mobility of virion proteins.

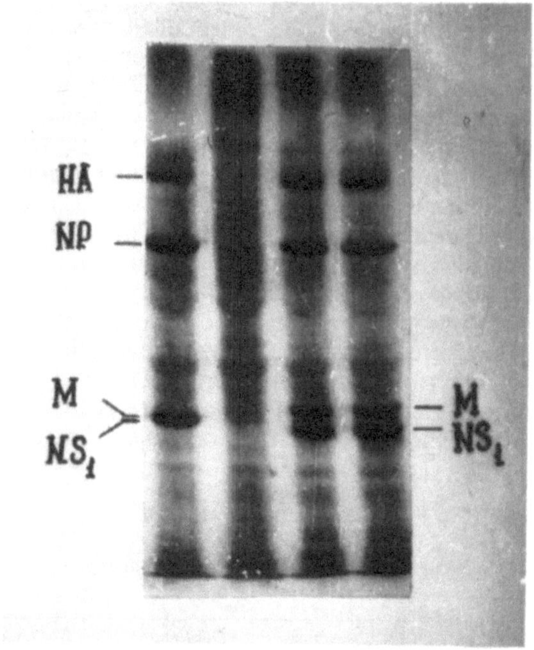

Figure 1 Electrophoresis of virus-specific proteins synthesized in MDCK cells after infection with original and persisting influenza virus variants; 15% polyacrylamide gel. ^{14}C-labelled protein radiography. Lane 1, virus isolated 6 months after infection of mice; lane 2, proteins from uninfected MDCK cells; lane 3, virus isolated 2 months after infection; lane 4, original A/Hong Kong/68 virus.

REFERENCES

Frolov, A.F. (1973) *Virusy i kantserogenez* (Viruses and Carcinogenesis). Kiev: Zdorovye (in Russian).

Frolov, A.F. (1981) *Etiologiya i patogenez grippa i ORZ* (The Etiology and Pathogenesis of Influenza and ARD). Leningrad; Meditsina (In Russian).

Frolov, A. F. and Gavrilov, S.V. (1981) *Virusy i virusnyye zabolevaniya* (Viruses and Viral Diseases), pp. 9, 68–73. Kiev: Zdorovye (In Russian).

Frolov, A.F., Rudenko, A.A., Markov, T.C. and Gavrilov, S.V. (1982) *Vrachebnoyedelo*, **1**, 110–113 (In Russian).

Frolov, A.F., Shcherbinskaya, A.M., Rubalko, S.L., Gavrilov, S.V. and Zatel, J.P. (1978) *Mikrobiol. Zh.*, **40**, 102–103 (In Russian).

Frolov, A.F., Shcherbinskaya, A.M. and Sklyanskaya, E.I. (1981) *Vopr. Virusol.* **5**, 544–547 (In Russian).

Frolov, A.F., Trinus, E.K., Rudenko, A.A., Dzyublik, I.V., Barchuk, V.F. and Sil'chenko, V.I. (1986) *Vrachebnoye delo*, **1**, 109–111 (In Russian).

Ritova, V.V. (1969) *Ostryye respiratornyye virusnyye infektsii u detei rannego vozrasta* (Acute Respiratory Viral Infections in Young Children). Moscow: Meditsina (In Russian).

Sov. Med. Rev. E. Virology Reviews, Vol. 3. 1990, pp. 37–60
Photocopying permitted by license only

PERSISTENCE OF INFLUENZA VIRUS IN DIFFERENT BIOLOGICAL SYSTEMS AND ITS POSSIBLE IMPLICATIONS

V. A. ZUYEZ, E. P. MIRCHINK AND
A. M. KHARITONOVA

N.F. Gamaleya Institute of Epidemiology and Microbiology, USSR Acadamy of Medical Sciences, Moscow

Abstract

The results of a study into the persistence of influenza virus, pursued by these authors since 1969 at the Gamaleya Institute of Epidemiology and Microbiology, USSR Academic of Medical Sciences, are briefly outlined.

The first stage of these experiments indicated the possibility of producing and continuously maintaining a latent influenzal infection in cell cultures previously considered insusceptible to influenza viruses. It was demonstrated that the latent viral persistence mechanism in such systems is conditioned by the disturbed assembly of viral ribonucleoprotein (RNP). A consistent pattern has been recognized, according to which either an acute or latent influenzal infection may be produced within the same "influenza virus-L 929 cells" system depending on the ratio of defective to standard particles in the infecting viral preparation and on the multiplicity of inoculation.

At the second experimental stage, a combined strategy was devised to demonstrate that the persistence of influenza virus in mammals may result from all known natural ways of its introduction: (1) recovery from the disease; (2) immunization with living virus; and (3) by vertical (mother–fetus) virus transmission. If the first two ways lead to a latent infection, the third route not infrequently gives rise to a slowly developing influenza infection in the progeny characterized by an unusually long incubation period, a slowly progressive course, peculiar damage to organs and tissues, and a fatal outcome.

1. RATIONALE

It took years of research to elucidate whether influenza virus could persist *in vivo*. The well-known ubiquity of influenza virus, the epidemic, and not infrequently pandemic, pattern of disease produced, and hence the inestimable harm to human health and enormous economic damage, all these factors taken together offered solid arguments that swayed our decision. To this it must be added that if it could be demonstrated that influenza virus was able to produce and maintain a latent infection *in vivo*, it would provide experimental evidence for the well-known supposition that latent influenza infection might be a means by which the influenza virus survives in an inter-epidemic period (Stuart-Harris, 1945; Andrews, 1969).

2. LATENT INFLUENZA INFECTION IN AN *IN VITRO* SYSTEM

In approaching these issues, one had first of all to find out whether influenza virus could basically maintain a latent form of the infectious process. It appeared that the task would be best accomplished through simple systems. Accordingly, we opted for the L929 cell culture, known formerly as not being susceptible to influenza viruses, or at least showing no signs of virus-specific degeneration upon infection. This feature seemed to be critical, since the persistence of influenza virus in cell cultures, that had been reported by that time (Gavrilov *et al.*, 1966, 1967, 1970), was accompanied by periods of cell degeneration and re-population, alternating on a cyclic basis, and thus qualified by the authors themselves as a chronic influenza infection *in vitro* (Gavrilov and Zuyev, 1973).

2.1 Carrier cells

Infection of L929 cell cultures with a routine preparation of the influenza A(H1N1) virus at a variable multiplicity of inoculation was never accompanied by the development of signs of virus-specific cell degeneration. By contrast, cells in cultures infected with the WSN or Waybridge strain of influenza virus turned out to be 1.5–2 times larger in size than the uninfected ones. The cultures were also distinguished by higher mitotic indexes and, accordingly, by a higher rate of cell proliferation (Table 1). The features noted were preserved by sub-inoculation over a period of up to two years (Timakov and Zuyev, 1970; Zuyev *et al.*, 1970; Timakov *et al.*, 1971, 1972).

The immunofluorescence analysis employed detected the specific fluorescence of the perinuclear membrane and cytoplasm in the infected

Table 1 Cell sizes and mitotic index values in L cell cultures infected with influenza virus.

L cell culture	*Cell size*[a]	*Mitotic index*[b]
Uninfected with influenza virus (control)	453.4	13 ± 0.9
Infected with influenza virus (WSN strain)	644.4	50 ± 3.4
Infected with influenza virus (Waybridge strain)	802.5	25.4 ± 2.6

[a] Mass of 60 cell projections in mg; [b] $M \pm SEM$, m; $n = 3$.

cultures, suggesting the presence in the cells of an influenza antigen. The number of antigen-containing cells in such cultures increased as they were passaged to come to 100% by 20 to 25 passages (Zuyev *et al.*, 1971; Azadova *et al.*, 1971).

The distinctions discovered between infected and control cell cultures allowed one to assume the presence of persisting influenza virus which before long was isolated from the nutrient medium of these infected L929 cell cultures provided test material was subjected to "blind" passages in a sensitive system, such as (Table 2) a chick embryo fibroblast culture (Timakov and Zuyev, 1970; Zuyev *et al.*, 1970, 1971; Azadova, *et al.*, 1971).

In this manner, direct evidence was obtained supporting the capacity of influenza virus to produce and maintain a latent infection under *in vitro* conditions (Zuyev, 1973; Mirchink and Zuyev, 1973).

2.2 Mechanism of viral persistence

Comprehensive research into the problem of latent, chronic and slow viral infections should probably focus on the mechanisms of viral persistence. That is why the next step in our work involved elucidating the mechanisms of maintaining a latent influenza infection in L929 cell cultures. In doing so, we proceeded from the well-known heterogeneity of viral populations (Huang and Batimore, 1970; Kantorovich-Prokudina and Semenova, 1982), including those of influenza virus (Kantorovich-Prokudina and Semenova, 1982; von Magnus, 1965).

On the basis of a comprehensive assessment, including determination of buoyant density in a sucrose concentration gradient, the morphology of virions, and their specific infectivity and interfering activity,

Table 2 Isolation of infectious influenza virus from latently infected L cell cultures.

Isolation characteristics and conditions	*Carrier cell passage number*						
	5th	9th	16th	17th	19th	23rd	24th
Period of L cell culture subsequent to infection with influenza virus (days)	40	71	129	132	153	185	192
Number of passages on chick firbroblast culture[a]	5	5	3	5	2	3	2
Isolated virus titre log TCD_{50}	0.75	0.5	0.25	0.25	0.5	0.25	0.25

[a] Number of passages needed for isolation of infectious influenza virus from the nutrient media of carrier cells.

populations of unpurified preparations of influenza virus were found to consist of three fractions. The first fraction with a buoyant density of 1.21–1.23 $g.mL^{-1}$ was enriched with standard particles, the second with a buoyant density of 1.18–1.19 $g.mL^{-1}$ contained approximately equal amounts of defective and standard virions, and the third fraction with a buoyant density of 1.16 $g.mL^{-1}$ was enriched with defective–interfering (DI) particles (Denisov *et al.*, 1975).

Infection of L929 cell cultures with different fractions of the influenza virus preparation soon proved to cause the development of various forms of the infectious process in these cultures. For instance, infection of cultures with influenza virus preparations enriched with standard particles leads, at a high multiplicity of inoculation, to the development of an acute infectious process characterized by the quick onset of a clearly defined cytopathic effect and the emergence of infectious progeny in the nutrient medium. At a low (less than 0.1) multiplicity of inoculation, "asymptomatic" viral persistence emerges in the cultures without any attendant signs of cell destruction despite the continued propagation in these cultures of the infectious influenza virus and its escape into the nutrient medium.

An entirely different picture appeared in the cells after their cultures were infected with influenza virus preparations enriched with DI particles. Arising in the cultures at any, even high (around 10) multiplicity of inoculation was a representative latent infection unaccompanied by any signs of cell degeneration and infectious virus production. Viral persistence was demonstrated only by the fluorescent antibody technique, which made it possible to detect the virus-specific antigen in such cells (Zuyev *et al.*, 1971; Azadova *et al.*, 1971). Moreover, the cell cultures thus infected were found to allow intensive synthesis of virus-specific RNA (Vishnichenko and Zuyev, 1972), and the cloning method showed that more than 60% of the population cells had influenza virus (Denisov *et al.*, 1975).

The overall evidence obtained led to the assumption that the mechanism of a latent influenza infection maintained in L cell cultures might be due to accumulation of DI particles, whose emergence in the process of viral persistence appears to impede normal virion assembly (Denisov *et al.*, 1975; Zuyev, *et al.*, 1974). Indeed, soon fractionation in a sucrose step gradient detected in the homogenates of latently infected cells a large number of drastically shortened RNP structures whose diameter closely corresponded to that of the influenza virus RNP. Electron microscopy revealed that in contrast to the multispiral influenza virus RNPs obtained under the same step-gradient conditions from the homogenates of chick embryo fibroblasts, where an acute influenza infection was produced, as was the case in permissive cells, the RNP from latently infected cells has, in most cases, a monospiral structure, rarely consisting of twin spirals or more. However, RNP structures whose length closely corresponds to that of influenza virus RNP reproducing in susceptible cells, were found in the same L929 cells, but infected with a standard influenza virus and, consequently, producing acute infection (Denisov *et al.*, 1975).

3. LATENT INFLUENZA INFECTION IN MICE

The second stage of our work on the persistence of influenza virus in different biological systems involved modelling and investigating the persistence of these viruses in mammals. With this in mind, use was made of both outbred and inbred mice, known to be highly sensitive to the influenza A virus.

Early attempts to demonstrate that influenza virus can persist in mice were early undertaken by Zuyev (1957), when the virus was isolated by blind passaging on chick embryos from lung tissue homogenates of mice immunized with living virus, three weeks after termination of the immunization cycle.

3.1 Isolation of persisting influenza virus

Later, in order to explore the possible long-term persistence of influenza virus in mammals, our laboratory designed the following combined method of recovering persisting influenza virus *in vivo*. Infectious virus was detectable in the blood and organ homogenates of mice infected intranasally with influenza virus within approximately 12 days. After that time, routine procedures failed to detect infectious virus in surviving animals *in vivo* (including the lungs) (Zuyev *et al.*, 1974; Pavlenko and Mirchink, 1977). In this connection, when analysing the factors favouring the masking of infectious virus in virus carriers *in vivo*, anti-influenza antibodies appeared to be the most likely candidates. Incidentally, these antibodies begin accumulating in the blood serum in sizeable amounts by only the 10th and 12th day. This is why the first step incorporated into the combined method was washing the test animal organs *in situ*.

Considering that the concentration of persisting virus is known to be invariably lower than its concentration in acute infection, it seemed reasonable, on virus isolation from organs, to destroy as completely as possible the tissue cells of a given organ, counting on the maximal production of persisting virus by virus-containing cells. With this in mind, we used synthetic corundum to destroy organ tissues, an abrasive material second only to diamond in terms of hardness. As can be seen

Table 3 Titres of influenza virus recovered from organ suspensions obtained using glass powder or synthetic corundum.

Organ	*Synthetic corundum*	*Glass powder*
Lung	6.25	4.75
Liver	3.25	2.5
Brain	2.5	2.5
Kidney	2.0	1.75
Spleen	2.0	2.0

from Table 3, the denser the test organ, the greater the efficiency of the synthetic corundum technique: the abrasive makes it possible to increase viral yield from the virus-containing cells of lung tissue nearly 50-fold.

Finally, the last component in the combined strategy for the isolation of persisting virus was the antiglobulin serum from mice. Serum application was based on the following consideration. Mice surviving after influenza virus infection were known to suffer pneumonia of variable severity. Therefore, while lung tissue homogenates are being prepared, pulmonary epithelium cells are subjected to destruction along with the plasmic (antibody-forming) cells that had survived the pneumonia. As seen from the data in Table 4, prewashing the lung free of circulating antibodies, combined with the destruction of lung tissue by synthetic corundum in the presence of antiglobulin serum, allows the isolation of infectious influenza virus from the lungs of apparently healthy mice five weeks after infection (Pavlenko and Mirchink, 1977; Timakov *et al.*, 1974).

3.2 Latent influenza infection in mice surviving after infection or immunized with living virus

The combined technique developed to isolate persisting influenza virus verified experimentally the possibility that a latent influenza infection might be produced and maintained among mammals in every known naturally occurring way, notably: after recovery from illness; by virtue of immunization with living virus; and from vertical (pregnant female–fetus) virus transmission.

As seen from the data in Table 5, mice surviving after infection produced persisting virus up to day 112 after its initial administration (infection). By contrast, mice immunized with living virus yielded infectious influenza virus up to day 35 only. However, despite the marked differences in the duration of viral persistence, latent influenza infections produced in the animals by the first two ways have several

Table 4 Effects of different treatments of lung tissue on the isolation of infectious influenza virus from the lungs of carrier mice.

Lung tissue homegenate dilution	*Synthetic corundum*	*Washing + synthetic corundum*	*Synthetic corundum + anti-globulin serum diluted*			*Washing + synthetic corundum + antiglobulin serum diluted*		
			1:16	1:32	1:64	1:16	1:32	1:64
Undiluted	0[a]	0	0	0	0	0	4/4[b]	2/4
10^{-1}	0	0	0	0	0	0	1/4	1/4
10^{-2}	0	0	0	0	0	0	1/4	0
10^{-3}	0	0	0	0	0	0	0	0

[a] The virus is absent in the test material.
[b] The fraction numerator represents the number of positive results, the denominator the number of samples tested.

Table 5 Isolation of persisting influenza virus from the lungs of mice surviving after infection or immunization with living virus.

	Days after virus inoculation										
Mouse group	14	21	28	35	42	56	70	84	98	112	126
Infected with influenza virus at a dose of 100 LD_{50}	70	70	50	n.t.	20	0	20	10	0	20	0
Immunized with influenza virus at a dose of 0.001 LD_{50}	50	0	0	40	0	0	0	0	0	n.t	n.t.

Values are percentages indicating the efficacy of isolation of infectious influenza virus on chick embryos; n.t. = not tested.

features in common: in both cases the animals looked practically healthy, persisting infectious virus was produced exclusively by lung tissue, and virus titres never reached one log EID_{50}. Hence, the conclusion that follows is that mammals produce and maintain a latent influenza infection after both recovery from illness and immunization with living virus (Zuyev *et al.*, 1974; Pavlenko and Mirchink, 1977; Timakov *et al.*, 1974).

4. SLOW INFLUENZA INFECTION IN MICE

Testing of the third pathway of modelling viral persistence in mammals – that is, its production by virus transmission from pregnant female to fetus (vertical route) revealed quite surprising phenomena. As can be seen from Table 6, all the organs tested (lung, liver, spleen, kidney and brain) and the blood of the progeny born of carrier females produced infectious influenza virus with titres between 1 and 3 log EID_{50}. A similar result was obtained when examining the offspring of female mice infected effectively with influenza virus through the placenta during pregnancy (Table 7).

Subsequent observations on the development of progeny born of both carrier females and females infected with influenza virus during pregnancy revealed that after three weeks part of the progeny showed signs of disease marked by progressively retarded growth and weight, motor incoordination, a waddling gait (the gait of a duck), and increasing emaciation; the disease was invariably fatal (Figure 1) (Mirchink *et al.*, 1977; Zuyev *et al.*, 1981, 1983).

Pathohistological changes discovered in such mice differed markedly from the principal signs of acute influenza infection in mice in general. In the first place, signs of inflammation were lacking in mice with a

Table 6 Isolation of infectious influenza virus from the organs of infant mice infected in *utero* by carrier mothers.

	Duration of carrier state in mothers (days)						
Tissue	63	77	91	98	105	119	133
Lung	3.0	2.5	2.5	3.0	1.0	0	0
Liver	2.0	2.0	1.0	n.t.	n.t.	0	0
Kidney	2.0	2.0	0	n.t.	n.t.	0	0
Spleen	2.0	1.0	1.0	n.t.	n.t.	1.0	0
Brain	2.5	2.5	1.0	n.t.	n.t.	1.0	0
Blood	n.t.	3.0	n.t.	2.5	1.0	1.0	0

Values are infectivity titres in log EID_{50} per mL) determined by infection of chick embryos; n.t. = not tested.

Table 7 Effectiveness of transplacental transmission of influenza virus in mice as a function of viral preparation and intranasal infection does (in LD_{50} or EID_{50}) of pregnant females.

	Viral preparation enriched with standard virions					*Viral preparation enriched with defective virions*			
	10^{-1} LD_{50}		10^{2} LD_{50}		10^{4} LD_{50}	$10^{6.5}$ EID_{50}		$10^{5.5}$ EID_{50}	$10^{4.5}$ EID_{50}
Tissue	Day 1	Day 2	Day 1	Day 2	Day 1	Day 1	Day 2	Day 1	Day 1
Female lung	4.0	6.25	7.5	8.5	8.75	5.5	6.0	7.0	6.5
Female blood	2.25	3.0	4.0	4.5	6.5	2.5	3.0	5.0	2.5
Placenta	0.25	1.5	2.5	3.0	4.5	1.5	1.5	1.5	1.5
Fetus	0	1.0	2.0	2.5	3.5	1.5	1.5	1.0	0.75

Values are titres of infectivity (in log EID_{50}) determined by infection of check embryos. Days are days after infection.

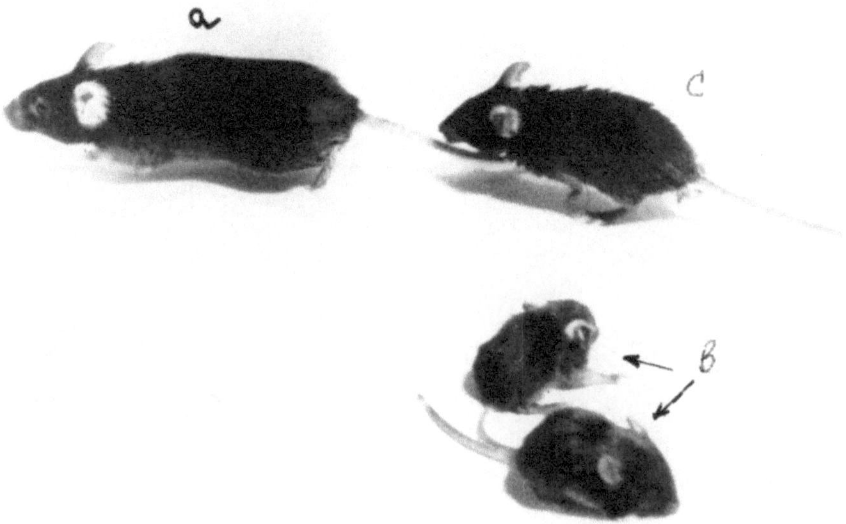

Figure 1 Part of the litter born of a C57BL/6 female: (a) infected with influenza virus during pregnancy; (b) individuals with pathological signs showing double retardation in growth (weight) compared with (c) the neighbouring sib.

congenital influenza infection, and the pathohistological changes were characterized by primary degenerative changes particularly pronounced in the central nervous system (CNS) and immunocompetent organs (Mirchink *et al.*, 1977; Zuyev *et al.*, 1981, 1983).

The relationships between the cortical and medullary substances in the thymus have been severely upset: the number of thymocytes in the cortical substance has dropped steeply, and the whole thymus looks homogenous; at times the cortical substance looks even lighter than the medulla ("histological inversion") (Figure 2).Histological examination reveals in the thymus the steeply reduced activity of the lysosomal enzymes – acid phosphatase and β-glucuronidase. The follicular structure is damaged in the spleen and lymph nodes, i.e. the red pulp follicles are barely detectable. In addition, there is also a severe decrease in the number of lymphocytes in the spleen because of their death.

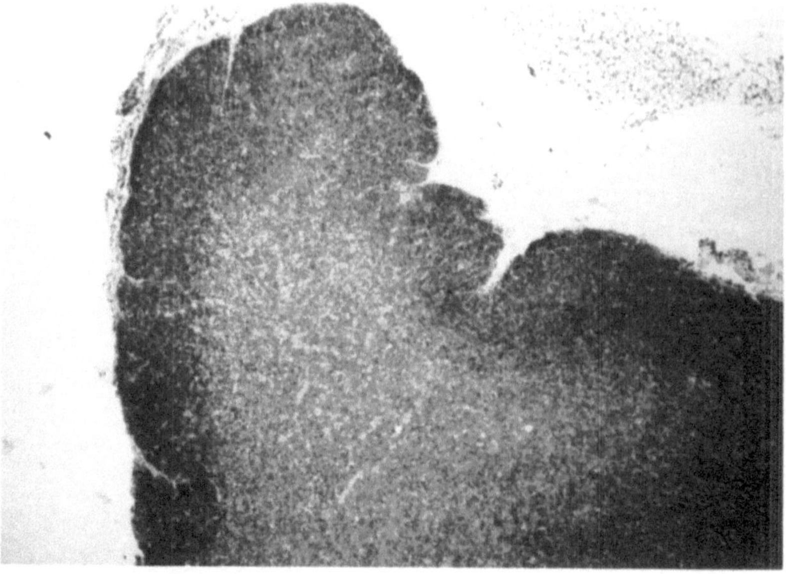

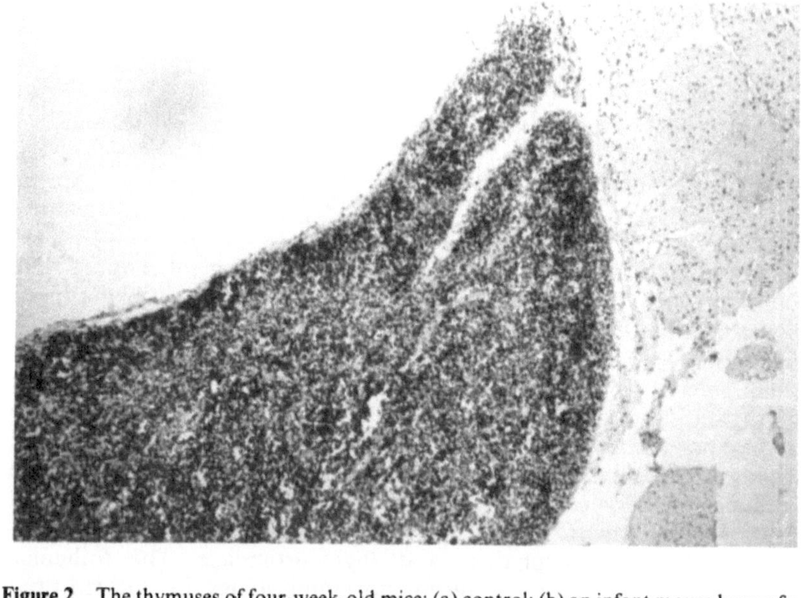

Figure 2 The thymuses of four-week-old mice: (a) control; (b) an infant mouse born of a female infected with influenza virus during pregnancy (a severe decrease in the number of thymocytes in the cortical substance is seen). (Hematoxylin and eosin; × 160.)

The liver and kidney exhibit signs of albuminoid degeneration of hepatocytes and of the epithelium of convoluted tubules in the kidney.

Changes in the CNS are manifested by an original complex of symptoms: namely neuronal hyperchromatosis, pyknosis, lysis and loss, as well as spongiosis in the white matter of the truncal parts of the cerebrum and the cerebellum, and gliosis (Figures 3–5). In some instances, there is a motoneuronal loss in the spinal cord (Figure 6), accompanied by pronounced signs of spongiosis in the white matter of the spinal cord (Figure 7). Distinct changes, such as pyknosis and hyperchromatosis, also occur in the supraoptic and paraventricular nuclei of the anterior hypothalamus and in the cells of arcuate and ventromedial nuclei representing what is known as the "adenohypophyseotropic zone" of the hypothalamus, which controls hormonopoiesis in the adenohypophysis (Voitkevich, 1967; Tarakaniv, 1968; Kakhana, 1968). The adenohypophysis itself shows a sharp reduction in the number of acidophils that, in turn, exhibit fewer secretory granules. The changes found in the organs regulating the hormonal activity of the endocrine glands, are attended by well-defined

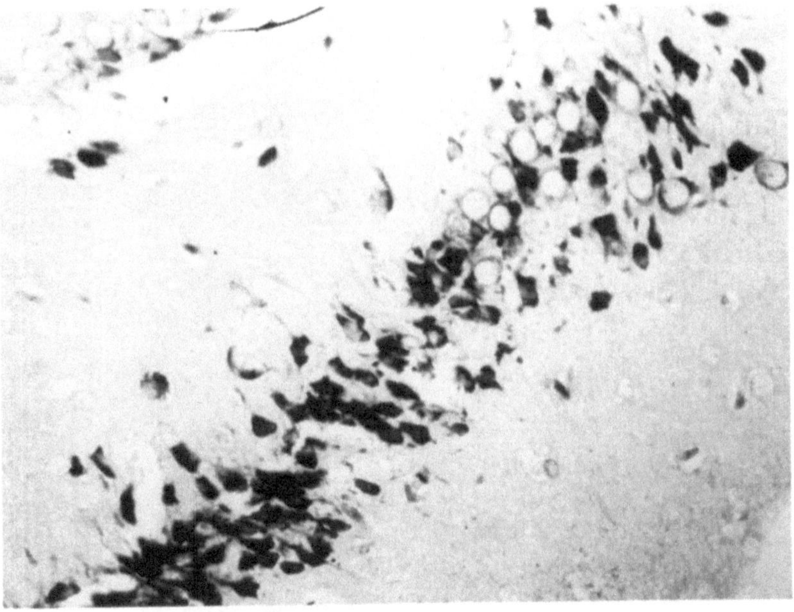

Figure 3 Neuronal pyknosis and hyperchromatosis in the hippocampus of the brain of an infant mouse born of a female infected with influenza virus during pregnancy. (Mayorova's stain; × 120.)

Figure 4 Neuronal loss in the pyramidal cell layer of the cerebral cortex of an infant mouse born of a female infected with influenza virus during pregnancy. (Mayorova's stain; × 200.)

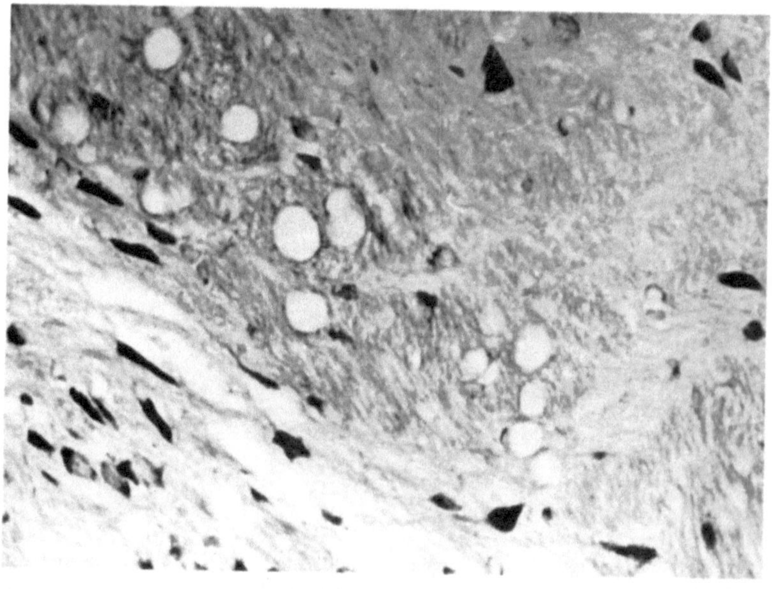

Figure 5 Spongiosis in the white matter of the truncal parts of the brain of an infant mouse born of a carrier female. (Mayorova's stain; × 180.)

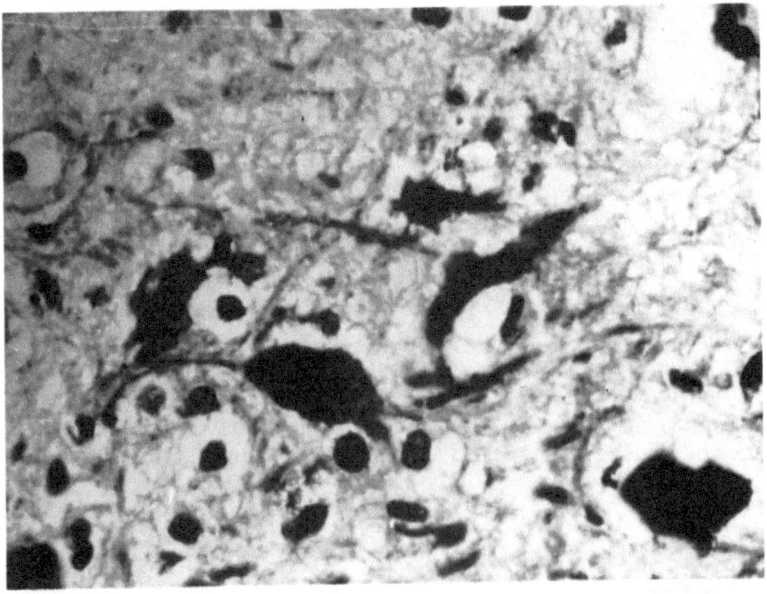

Figure 6 Motoneuronal loss in the anterior horns (thoracic part) of the spinal cord of an infant mouse born of a carrier female. (Nissle's stain; × 450.)

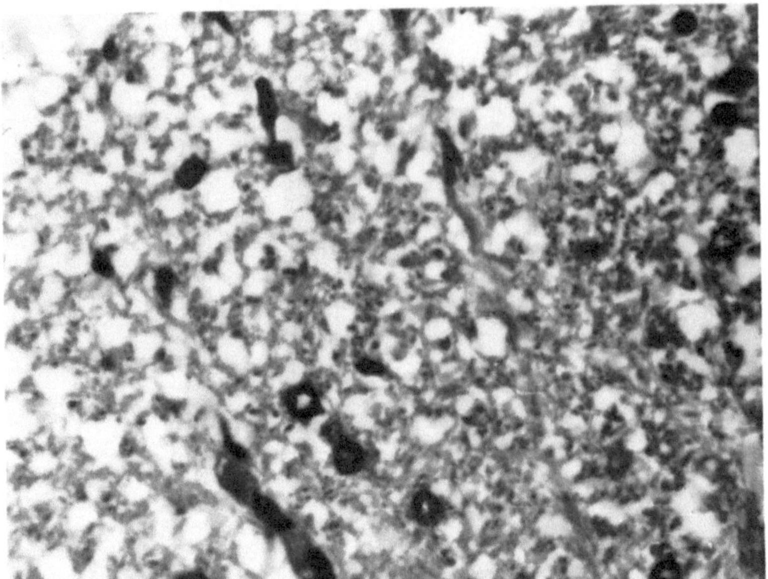

Figure 7 Spongiosis in the white matter of the anterior horns of the spinal cord of an infant mouse born of a female infected with influenza virus during pregnancy. (Hematoxylin and eosin; × 300.)

changes in the sexual and thyroid glands and in the adrenal glands known as "target glands".

The foregoing degenerative changes in immunocompetent organs, are combined with a severe suppression of the immune response in the animals (Table 8). As one can see from Table 8, in mice infected with influenza virus *in utero* the degree of inhibition of the immune response in infancy turns out to vary and is far more pronounced in cases where pregnant females were infected with the virus enriched with standard virions. In this case antibody-forming cells (AFC) and rosette-forming cells (RFC) accounted for 28.8 and 27.8%, respectively, compared with controls. The steep reduction in the number of immunocompetent cells is supplemented by inhibition of their functional activity, indicated in the lymphocyte blast-transformation reaction to both non-specific T and B mitogens and a virus-specific antigen provided by UV-inactivated influenza virus (Figures 8 and 9). The immunological state of animals

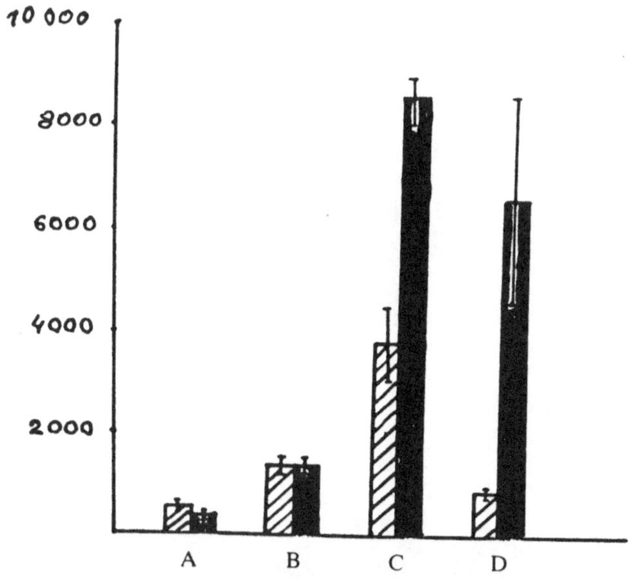

Figure 8 Spleen lymphocyte blast-transformation reaction to T and B mitogens in control infant mice (black bars) and in dwarf infant mice with a slowly progressing influenza infection (cross-hatched bars):
Ordinate: activity (impulse. min – ¹);
Abscissa: A = phytohemagglutinin; B = without mitogen;
C = concanavalin A; D = dextran sulfate.

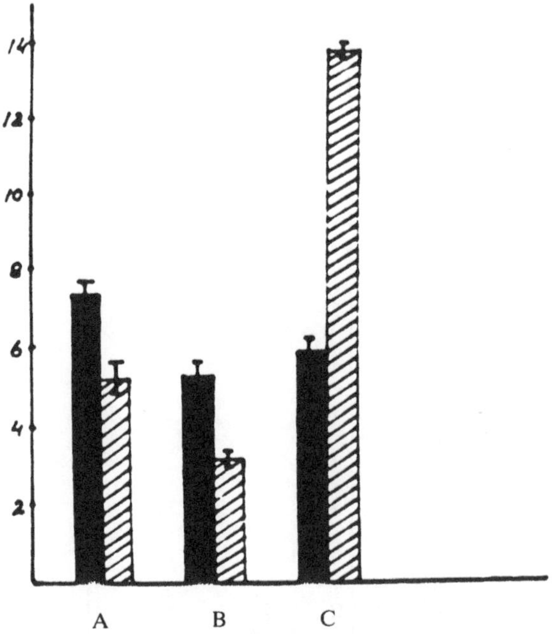

Figure 9 Spleen lymphocyte blast-transformation reaction to the influenza virus antigen in control and immune infant mice as well as in dwarf infant mice affected by slow influenza infection: black bars, without antigen; cross-hatched bars, with antigen added.
Ordinate: activity (impulse. min $-^{1}$ × 10 $-^{3}$);
Abscissa: A = control; B = dwarf; C = immune mouse.

with the slowly progressing process is thus characterized by demonstrable immunodepression (Mirchink *et al.*, 1984; Zuyev, 1988).

5. CONCLUSIONS

These data reaffirm the proposition that the forms and consequences of viral persistence may vary with conditions and the route by which it is established. This idea is apparently of universal importance, the discussed feature of viral persistence being shown in *in vitro* and *in vivo* systems alike.

A review of the experimental evidence presented in the last section suggests that intrauterine infection of mouse fetuses with influenza virus (either from carrier females or from females infected during pregnancy)

Table 8 Primary immune response to sheep erythrocytes in infant mice born of females infected with influenza virus during pregnancy.

	Antibody-forming cells (AFC)			*Rosette-forming cells (RFC)*		
Group of infant mice	*per 10^6 cells*	*per spleen*	%	*per 10^3 cells*	*per spleen* ($\times 10^{-5}$)	%
Born of uninfected females (control)	29.0 ± 6.7	6235 ± 1780	100	6.3 ± 0.84	13.5 ± 1.7	100
Born of females infected with influenza virus enriched with standard virions	12.4 ± 2.3	1793 ± 221	28.8	2.6 ± 0.4	3.76 ± 0.6	27.8
Born of females infected with influenza virus enriched with defective virions	23.3 ± 6.6	3704 ± 1540	59.4	4.8 ± 0.8	7.57 ± 2.4	56.1

Values are M ± SEM; n(number of measurements) = 30.

may cause part of the progeny to develop a disease characterized by:

1. an extended (three-week) incubation period;
2. a slowly progressive course;
3. lack of inflammation signs despite the devlopment of primary degenerative changes, particularly pronounced in the CNS and immunocompetent organs; and
4. a fatal outcome.

These characteristics fully agree with the principal traits typical of slow viral infection (Sigurdsson, 1954) which are: a long incubation period; a slowly progressive course; unusual lesions of organs and tissues; and a fatal outcome. Consequently, these findings can be thought of as systematic evidence for the capacity of influenza virus to induce, upon intrauterine infection of fetuses, the development of slow influenza infection in offspring (Zuyev *et al.*, 1985; Zuyev, 1988).

The literature reveals many reports on standard viruses (such as rubella, lymphocytic choriomeningitis, herpes and others) that produce a slow infection in progeny exclusively on condition that the fetuses had been previously infected intrauterinely (Rawls, 1974; Cole and Nathanson, 1974), implying, naturally, the ability of such viruses to cross the placental barrier. The ability of influenza virus to be transmitted through the placenta is well known, being first demonstrated in humans (Yawn *et al.*, 1971; Ruben *et al.*, 1975; Ritova, 1976) and further corroborated by numerous experiments in animals (mice) (Ritova, 1976; Seim *et al.*, 1960; Korsatinya and Bakhutashvili, 1974; Flamm, 1962). In this context, it should be recalled that an increase in congenital pathology after influenza epidemics – a fact noticed long ago – is not uncommonly combined with higher maternal lethality, abortions, and premature deliveries (Ritova, 1976; Flamm, 1962). Various CNS lesions in influenza, varying in location and clinical manifestations, warrant particular attention (Yakimenko *et al.*, 1975). Finally, the ability of influenza virus to cause pronounced immunodepression *in vivo* is still another property of the virus, which is important in elucidating the mechanisms by which different organs and tissues are affected and, consequently, in gaining some insight into the conditions under which various forms of influenza infection develop (Yakimenko *et al.*, 1975; Kiselyova *et al.*, 1982; Gardner *et al.*, 1973). Thus, there are many grounds for the development of the slow form of influenza in special conditions of infection (Zuyev, 1988).

In summary, we have presented several features that distinguish the above-described slow influenza infection in mice from spongiform polioencephalomyelopathy of mice produced by retrovirus type C

(Gardner *et al.*, 1973; Brooks *et al.*, 1980; Gardner, 1985). They are:

1. epidemiological – in the slow influenza the highest incidence reaches 15%, in the retrovirus type C 85%;
2. clinical – in slow influenza infection the sick animals have no paralyses; and
3. pathogenetical – in slow influenza infection, immunocompetent organs, including the thymus, are damaged, and retroviral particles are absent in the suspensions of muscular tissue taken from the sick animals.

It may be added that a drop in the expression of protein P30 – the major internal protein of endogenous retroviruses type C – has recently been demonstrated in mice of different lines, which maintained influenza virus persistence for up to three months (Savtsova *et al.*, 1984).

REFERENCES

Andrews, C. (1969) Influenza. In *Yestestvennaya istoriya virusov* (The natural history of viruses), p.99. Moscow: Mir publishers (In Russian).

Azadova, N.B., Peters, V.V. and Zuyev, V.A. (1971) Discovery and study of the dynamic accumulation of influenza antigen in latent infected cell cultures. *Vopr. med. virusol.*, **2**, 50–51 (In Russian).

Brooks, B.R., Swarz, J.R. and Johnson, R.T. (1980) Spongiform polioencephalomyelopathy caused by a murine retrovirus. 1. Pathogenesis of infection in newborn mice. *Lab. Invest.*, **43**, 480–486.

Cole G.A. and Nathanson, N. (1974) Lymphocytic choriomeningitis. *Progr. Med. Virol.*, **18**, 94–110.

Denisov L.A., Manykin, A.A., Klimenko, S.M., Zuyev V.A. and Timakov V.D. (1975) Role of defective virus particles in mechanism formation of latent influenza infection. *Vestn. Akad. med. Nauk SSSR*, **12**, 3–7 (In Russian).

Flamm, G. (1962) Influenza. In *Prenatalnyye infektsii cheloveka* (Die pränatalen Infektionen des Menschen), pp. 79–80. Moscow: Medgiz publishers (In Russian)

Gardner, M.B. (1985) Retroviral spongiform polioencephalomyelopathy. *Rev. Infect. Dis.*, **7**, 99–110.

Gardner, M.B., Henderson, B.E., Officer, J.E., Rongey, R.W., Parker, J.C., Oliver, C., Estes, J.D. and Huebner, R.J. (1973) A spontaneous lower motor neuron disease apparently caused by indigenous type C RNA virus in wild mice. *J. Natl. Cancer Inst.*, **51**, 1243.

Gavrilov, V.I. and Zuyev, V.A. (1973) In search of latent form of influenza. *Priroda*, **9**, 78–85 (In Russian).

Gavrilov, V.I., Solovyeva, A.I. and Vyalushkina, S.D. (1970) Development of a model of chronic influenza infection of a cell line derived from nonprimate animal. *Vopr. virusol.*, **1**, 14–20 (In Russian).

Gavrilov, V.I., Solovyeva, A.I., Blumkin, V.N. and Azadova, N.B. (1966) Acute and subchronic infection of human diploid cell cultures by influenza A/PR8 virus. *Vopr. virusol*, **4**, 456–463 (In Russian).

Gavrilov, V.I., Solovyeva, A.I., Zhdanov, V.M. and Blumkin, V.N. (1967) A phenomenon of peculiar "masking" of influenza A virus (PR8) in cultures of serially passaged diploid cells of human embryo lung. *Vopr. virusol.*, **2**, 150–160 (In Russian).

Huang, A.S. and Batimore, D. (1970) Defective viral particles and viral disease processes *Nature*, **226**, 325–327.

Kantorovich-Prokudina, E.N. and Semenova, N.P. (1982) Defective interfering virus particles. *Uspek. sovrem. biol.*, **93**, 338–351 (In Russian).

Kakhana, M.S. (1968) Basic regulation's mechanisms of the hormonal functions. In *Patofiziologiya endokrinnoi systemy* (The pathophysiology of the endocrine system), pp. 5–26, Moscow, Meditsina publishers (In Russian).

Kiselyova, I.V., Guseva, V.M. and Ilyenko, V.I. (1982) The effect of remantadine on the immunosupressive action of influenza virus. *Vopr. virusol.*, **4**, 437–440 (In Russian).

Kobzon, N. (1964) Effect of the influenza virus on the pregnancy. *Med. zhourn. Uzbekistana*, **3**, 28–35 (In Russian).

Magnus von P. (1965) The in ovo production of incomplete virus by B/Lee and A/PR8 influenza viruses. *Arch. ges. Virusforsch.*, **17**, 414–423.

Mirchink, E.P. and Zuyev, V.A. (1973) Stimulation of cell multiplication with standard influenza virus and the virus isolated from latently infected cells. *Vestnik Akad. Med. Nauk SSSR*, **2**, 43–46 (In Russian).

Mirchink, E.P., Zuyev, V.A. and Miroshina, E.K. (1984) Slow influenza infection developing in the progeny of female mice infected during pregnancy. *Vopr. virusol.*, **1**, 32–35 (In Russian).

Mirchink, E.P., Zuyev, V.A., Pavlenko, R.G. and Kharitonova, A.M. (1977) Vertical transmission of the persisting influenza virus in mice and connected with that pathology in progeny. In *Abstracts of Papers of the 16th All-Union Congress of Microbiologists and Epidemiologists* (Ulianovsk, 4–7 October, 1977), ed. by P.N. Burgasov and V.M. Zhdanov, **3**, 52–54 (In Russian).

Mirchink, E.P., Pronin, A.V., Barteneva, N.S., Sanin, A.V., Khorobrykh, V.V. and Zuyev, V.A. (1984) Immunological analysis of pathology developing in mice as a result of intrauterine infection with influenza virus. *Vopr. virusol.*, **2**, 162–166 (In Russian).

Pavlenko, R.G. and Mirchink, E.P. (1977) Persistence of influenza virus in the organism of sensitive animals infected with different doses of the virus. *Vestnik Akad. Med. Nauk SSSR*, **7**, 16–21 (In Russian).

Rawls, W.E. (1974) Viral persistence in congenital rubella. *Progr. Med. Vorol.*, **18**, 273–288.

Ritova, V.V. (1976) Perinatal infection. In *Rol virusov v perinatalnoi i postnatalnoi patologii cheloveka* (The role of viruses in the perinatal and postnatal pathology of man), pp. 30–39, 60–82 and 91–106. Moscow: Meditsina (In Russian).

Rogers, S.C. (1972) Influenza and congenital abnormalities. *Lancet*, **1**, 261.

Savtsova, E.D., Orlovsky, A.A., Struk, V.I. *et al.* (1984) Change expression of some cell's proteins attached to chronic influenza infection of mice. *Dokl. Akad. Nauk USSR*, series B, **10**, 78–80 (In Russian).

Seim, R.A., Ly, H., Imagawa, D.T. and Adams, J.M. (1960) Influenza virus infections in pregnant mice. *J. Neuropathol. Exp. Neurol.*, **19**, 125–129.

Sigurdsson, B. (1954) Rida, a chronic encephalitis of sheep with general remarks on infections which develop slowly and some of their special characteristics. *Brit. vet. J.*, **110**, 341–354.

Stuart-Harris, C.H. (1945) Influenza epidemics and the influenza viruses. *Brit. Med. J.*, **1**, 209.

Tarakaniv, E.I. (1968) *Neirosekretsiya v norme i patologii* (Normal and pathological neurosecretion), pp. 3–219, Moscow, Meditsina publishers (In Russian).

Timakov, V.D. and Zuyev, V.A. (1970) Latent and chronic viral infections. Present-day state and key aspects of investigations. *Vestnik Akad. Med. Nauk SSSR*, **10**, 3–12 (In Russian).

Timakov, V.D., Zuyev, V.A. and Peters, V.V. (1971) Latent infection of cell cultures insensitive to the cytopathic effect of virus. Communication 1. Reaction of L cell cultures to infection with influenza type A virus. *Vopr. virusol.*, **16**, 281–285 (In Russian).

Timakov, V.D., Zuyev, V.A., Mirchink, E.P. and Peters, V.V. (1972) Alterations of L-cell cultures infected by influenza viruses. *Arch. Ges. Virusforsch.*, **37**, 279–281.

Timakov, V.D., Zuyev, V.A., Popenenkova, Z.A. and Romanovskaya M.G. (1974) Formation of the latent influenza infection in white mice. *Dokl. Akad. Nauk SSSR*, **215**, 1231–1232 (In Russian).

Vishnichenko, V.K. and Zuyev, V.A. (1972) Synthesis of virus-specific molecules in the L-cells attached to formation of latent infection by fowl plague virus. In *Molekulyarnaya biologiya virusov* (The molecular biology of viruses), pp. 64–65. Moscow. D.I. Ivanovsky Institute for Virology publishers (In Russian).

Voitkevich, A.A. (1967) *Neirosekretsiya* (Neurosecretion), pp. 3–250, Leningrad, Meditsina publishers (In Russian).

Yakimenko, L.V., Rybalko, S.L., Savtsova, Z.D., Verkhatsky, P.P. and Umansky, Yu.A. (1975) The immunodepressive effect of influenza virus. *Vopr. virusol.*, **1**, 96–98 (In Russian).

Yawn, D.H., Pyeatte, J.C., Joseph, J.M. *et al.*, (1971) Transplacental transfer of influenza virus. *J. Am. Med. Assoc.*, **216**, 1022–1023.

Zuyev, V.A. (1957) The significance of cellular reactions in immunity in experimental influenza. *Vopr. virusol.*, **3**, 176–179 (In Russian).

Zuyev, V.A. (1973) Latent influenza infection – a new type of virus persistence. *Vestnik Akad. Med. Nauk SSSR*, **2**, 3–15 (In Russian).

Zuyev, V.A. (1988) Slow influenza infection of mice. In *Medlennie virusnyye infektsii cheloveka i zhivotnich* (Slow virus infections in man and animals), pp. 115–136. Moscow, Meditsina publishers (In Russian).

Zuyev, V.A., Mirchink, E.P. and Kharitonova, A.M. (1985) Experimental evidence for the capacity of persistent influenza virus of inducing slow infection. *Vestnik Akad. Med. Nauk SSSR*, **3**, 26–31 (In Russian).

Zuyev, V.A., Mirchink, E.P. and Kharitonova, A.M. (1983) Experimental slow influenza infection in mice. *Vopr. virusol.*, **1**, 24–29 (In Russian).

Zuyev, V.A., Peters, V.V. and Azadova, N.B. (1971) Latent infection of cell cultures insusceptible to the virus cytopathic effect. *Vopr. virusol.*, **6**, 713–717 (In Russian).

Zuyev, V.A., Mirchink, E.P., Kharitonova, A.M., Denisov, L.A. and Belyayev, D.L. (1981) Experimental slow influenza infection in mice. In *Virusy i virusnyye infektsii cheloveka* (Viruses and viral infections in man), pp. 200–201. Moscow. Institute of poliomyelitis and viral encephalitises USSR Acad. Med. Sci. publishers (In Russian).

Zuyev, V.A., Mirchink, E.P., Vishnichenko, V.K., Azadova, N.B. and Mentkevich, L.M. (1974) The nature and mechanism of virus persistence in cell cultures with latent infection with avian influenza virus. *Vopr. virusol.*, **4**, 450–454 (In Russian).

Zuyev, V.A., Peters, V.V., Mirchink, E.P., Azadova, N.B. and Blinova, L.I. (1970) Latent infection of cell cultures by myxoviruses. In *Proceedings of the 15th All-Union Congress of Epidemiologists, Microbiologists, and Infectionists*, pp. 278–279. Moscow. Ed. by P.N. Burgasov (In Russian).

Zuyev, V.A., Pavlenko, R.G., Mirchink, E.P., Kharitonova, A.M., Belyayev, D.L. and Denisov, L.A. (1981) Possible ways of modelling latent influenza infection in mice. *Vopr. virusol.*, **3**, 280–285 (In Russian).

Zuyev, V.A., Popenenkova, Z.A., Romanovskaya, M.G., Mirchink, E.P., Pavlenko, R.G., Denisov, L.A. and Maslennikov, G.A. (1974) Use of mice for modelling of latent influenza infection in mammals. In *Laboratornyye zhivotnyye v meditsinskikh issledovaniyakh* (Laboratory animals in medical research), pp. 86–88, Moscow. Scientific laboratory of experimental biology models publishers.

Sov. Med. Rev. E. Virology Reviews, Vol. 3. 1990, pp. 61–78
Photocopying permitted by license only

CHARACTERISTICS OF THE CURRENT ETIOLOGICAL PATTERN OF ACUTE RESPIRATORY DISEASES IN THE USSR

T.I. YURLOVA, N.B. RUMEL, V.A. FURAYEVA and V.P. LYKOV

All-Union Research Institute of Influenza, USSR Ministry of Health, Leningrad

Abstract

Ten years of observation of the dynamics of the etiological pattern of acute respiratory diseases (ARD) in the USSR have brought to light some etio-epidemiological peculiarities of different ARD nosological forms and defined epidemically relevant types of adenoviruses and picornaviruses and the dynamics of their succession, depending on the period of observation throughout the country in general, and in selected climatic and geographical zones in particular. Two subtypes of adenovirus serotype 3, distinguished by cytotoxic activity in tissue cultures, have been identified.

The polymorphism of naturally occurring populations of adenoviruses (serotype 3), the respiratory syncytial virus (RSV), and the parainfluenza virus (serotype 3) has been determined by a number of antigenic, biological, and molecular and biological properties. A correlation between the biological profile of circulating viral populations (RSV and parainfluenza virus) and their epidemic potentials among institutionalized children has been established.

1. INTRODUCTION

Much information is available concerning the proportion of selected nosological forms and the dynamics of the etiological pattern of ARD as a function of observation periods (seasonal recurrence) and the age of subjects (age-related dynamics).

However, the problems of specific epidemiology, concerned with the circulation and variability of selected agent types, their strains in different climatic and geographical zones, and their periodic succession during varying observation periods, are as yet not fully understood. Finding solutions to these problems is of great utility to health care which, among other things, involves the development and practical application of viral vaccines prepared from strains with the current antigenic structure and specific genotypical or phenotypical profiles. Studies warranting special attention in this respect are those into the molecular epidemiology of adenovirus infections (Wadell, 1984), underlaid by evidence of epidemically relevant adenovirus serotypes for different regions of the world and in particular their genome's structure and variability. However, the periodic succession of adenovirus populations in different areas, seasons and age groups is still not clearly understood.

The antigenic heterogeneity of natural populations is intrinsic in parainfluenza viruses, serotypes 1 and 2 (Rybinskaya, 1975, 1976), and the respiratory syncytial virus (Larry, 1985; Mufson *et al.*, 1985). There is, however, hardly any information on a possible relationship between the level of variability of respiratory viruses and their epidemic activity, or between particular immunogenetic factors in the body and the prevalence of strains with specific genotypical and phenotypical properties. For these reasons our research has been focused for many years on the issue cited above.

2. BASIC RULES GOVERNING THE CIRCULATION OF RESPIRATORY VIRUSES IN THE USSR

In order to make the principal pattern of the nationwide circulation of ARD viruses clear, it is necessary first to set out the general and specific mechanisms governing the prevalence of selected nosological ARD forms.

The general mechanisms result from the basic background, the development of infection: that is, the prevalence rates at different seasons and in different population age groups; associative links in the dynamics of circulation of individual ARD agents during different observation periods; and lastly, the effect of the climatic and geographical features of different areas on the prevalence rate of ARD nosological forms.

Specific mechanisms arise from the established etiological pattern of distinct ARD forms (adenovirus, picornavirus, parainfluenza virus, etc.), the identified epidemically relevant types in the genesis of ARDs, and the dynamics of their succession in different observation periods, climatic and geographic regions, and population age groups.

To define the overall circulation pattern of ARD viruses, we reviewed information on virological and serological diagnosis of ARDs, over a 10-year observation period, obtained from different regions, including cities and towns in Siberia, the Soviet Far East, the Ukraine, the Urals, and those situated at the middle and northern latitudes of the European part of the USSR. In an effort to harmonize tests, serological diagnosis was made with the use of standard diagnostic kits, and virological diagnosis – virus identification and biological characterization – was carried out exclusively at the Research Institute of Influenza, USSR Ministry of Health. Over the 10-year period we reviewed more than half a million serological tests, carried out in 32 cities of the USSR, and over 10 000 virological tests.

While defining the spectrum of viruses circulating in the Soviet Union, we found that adenoviruses were prevalent among non-influenza viruses, their proportion being the greatest between 1984 and 1985. The prevalence of adenovirus illnesses followed a definite recurrence pattern, peaks being reported every third to fourth year of observation.

The proportion of respiratory syncytial viral (RSV) infection varied from 7.1 to 16%, recurring every two to three years.

No consistent recurrence pattern of the prevalence rate of picornavirus illnesses, caused by ECHO and type B Coxsackie viruses, emerged. It should be noted that in recent years of observation the proportion of ARDs of picornaviral nature has almost doubled.

The proportion of parainfluenza infection ranged between 9.9 and 12.2%, being only slightly inferior to that of RSV infections.

The seasonal factor, or the seasonal complex (climate, weather) had a material effect on the incidence rate; the winter seasonal pattern is not peculiar only to influenza. Among non-influenza illnesses, it is most pronounced in parainfluenza and RSV infections, and least in adenovirus infection. Parainfluenza, RSV and adenovirus infections are characterized by fall–winter (not uncommonly summer), spring–winter and spring–fall–winter seasonal recurrence patterns, respectively. In contrast, the acute respiratory infections of picornaviral etiology show a well-defined summer seasonal pattern. These illnesses usually occur in spring and summer months, being local outbreaks or sporadic in character.

Immunological factors determine the age dynamics of incidence and the recurrence of the rise and fall of ARDs at different seasons. The age-incidence pattern represents both the age immunological pattern of the population and its epidemiological activity.

RSV infection was prevalent among preschool children, influenza, parainfluenza and mycoplasmal infections among schoolchildren, and an adenovirus infection among adults. This relationship was seen throughout the year.

In calculating extensive mean annual figures for incidence (per 1000 persons in each age group) between 1970 and 1984, the proportion of adenoviruses, RS viruses, parainfluenza viruses, and *Mycoplasma pneumoniae* were found to have increased markedly particularly in preschool children. The trend toward the higher prevalence rate of mixed infections in school-age children had begun to show (Figure 1).

In defining the associative linkage in the circulation dynamics between influenza viruses and other ARDs in the USSR, the influenza virus was found to have a definite influence on the revival of adenovirus (caused by latent types) and herpetic (herpes simplex 1) infections. Thus, for instance, the proportion of these infections increased dramatically during the period of epidemics; this pattern was to be seen up until 1980. The proportion of influenza has of late increased in the inter-epidemic period. A shift in the temporal peak in the circulation intensity of influenza viruses has markedly affected the circulation pattern of adenoviruses of latent serotype and herpes simplex viruses. For example, in 1981 the maximum isolation of adenoviruses (latent types) and herpes viruses occurred in a period of seasonal rise in morbidity; whereas in 1982 it was in the summer period.

Influenza viruses seem to exert a powerful immunodepressive action on the body, resulting in activation of latent viruses partially integrated in the cell's genetic apparatus.

To summarize, the increased proportion of influenza in an inter-epidemic period has brought about a certain shift in the temporal peak in the prevalence of other infections. Understanding of this is vital in preventing these infections.

The natural, climatic, and geographical peculiarities of different areas had a considerable impact on the prevalence rate of parainfluenza, adenovirus, and RSV infections (Figure 2). By way of example, parainfluenza infection prevailed in areas of the Ukraine and in the north of the Russian Federation, adenovirus infections in the Urals, and RSV infection in the Ukraine. On the other hand, in north-western areas of the Russian Federation, there has been no marked increase in the proportion of these infections of late.

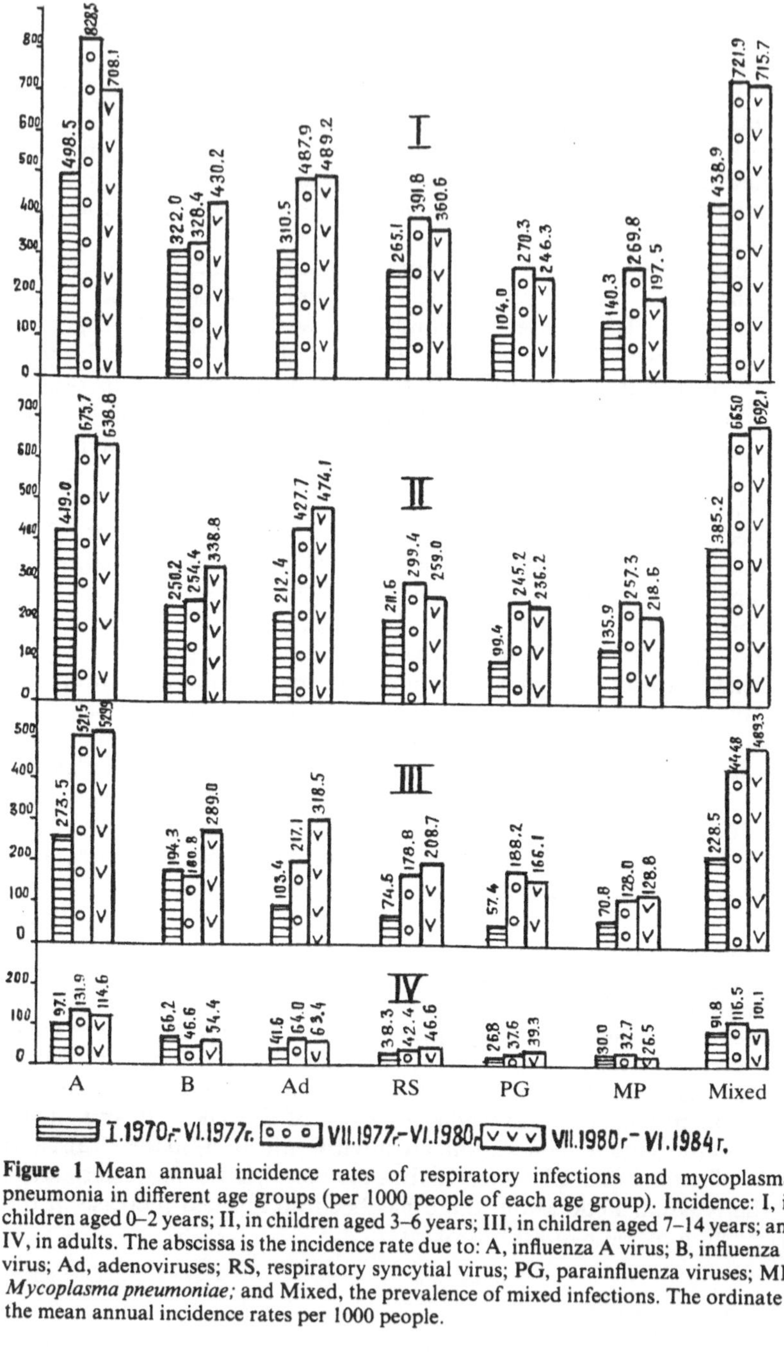

Figure 1 Mean annual incidence rates of respiratory infections and mycoplasmal pneumonia in different age groups (per 1000 people of each age group). Incidence: I, in children aged 0–2 years; II, in children aged 3–6 years; III, in children aged 7–14 years; and IV, in adults. The abscissa is the incidence rate due to: A, influenza A virus; B, influenza B virus; Ad, adenoviruses; RS, respiratory syncytial virus; PG, parainfluenza viruses; MP, *Mycoplasma pneumoniae;* and Mixed, the prevalence of mixed infections. The ordinate is the mean annual incidence rates per 1000 people.

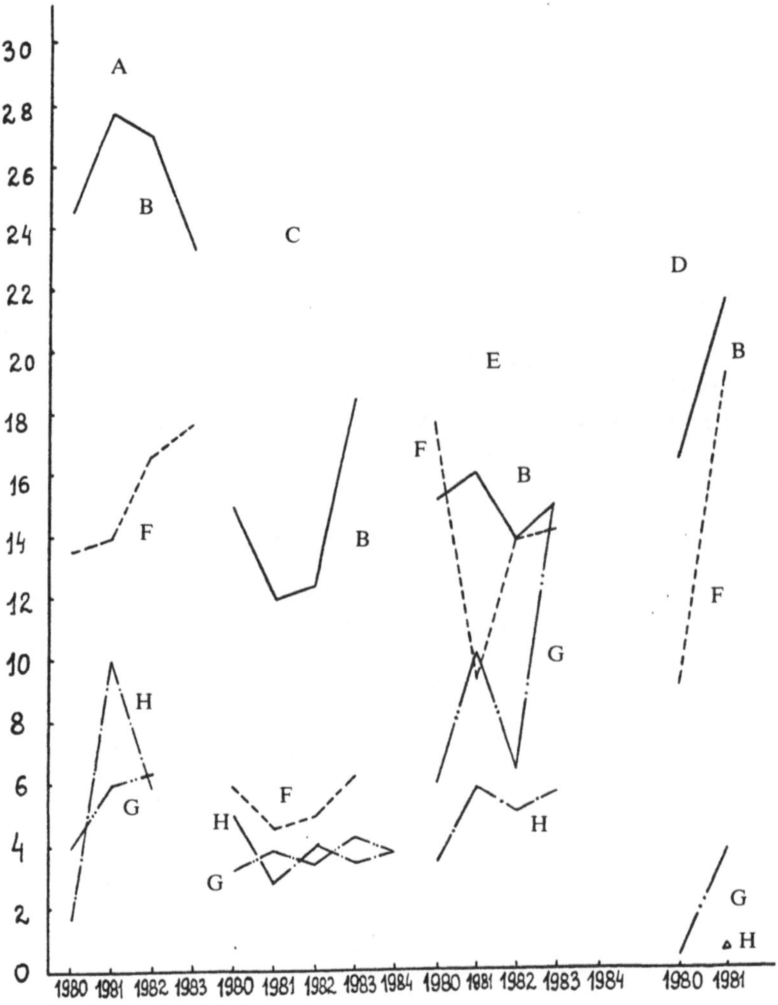

Figure 2 The effect of natural territorial features on the prevalence rate of different ARD nosological groups. The abscissa is the years of observation. The ordinate is the percentage identification of seroconversions to selected ARD agents, such as adenoviruses (adeno), respiratory syncytial (RS) virus, and parainfluenza (PG) viruses.

A, the Ukraine; B, influenza; C, the northwestern Russian Federation; D, the northern Russian Federation; E, the Urals; F, parainfluenza; G, adenoviruses; H, respiratory syncytial.

2.1 Adenoviruses

Individual types of adenovirus vary widely in importance for the genesis of ARDs. The so-called epidemic types of adenoviruses (types 3, 7, 21 and 14) had earlier been dominant in the etiological pattern of adenovirus diseases. Current adenovirus infections are characterized by a greater proportion of what is known as adenovirus latent types in the development of ARDs, compared to average isolation rates for these agents in previous years. For instance, adenovirus serotypes 3 and 7 had been virtually the only ones of epidemic relevance up to 1981–1982. Recent years have seen a tendency for these hitherto important adenovirus serotypes to be replaced; latent types (2, 6 and 1,5) have now acquired epidemic significance (Table 1).

In studying the dynamics of the circulation of adenovirus serotypes 3 and 7 in the USSR over the period of ten years, we found that the intensity of development of the epidemic process due to these viruses varied with the period of survey: it was comparatively high before 1979–1980 and low between 1984 and 1985 (Figure 3). By contrast, beginning in 1981–1982, the isolation rate of adenovirus latent types (2 and 6) has increased significantly, pointing to an intensified epidemic process due to these agents during recent years.

The isolation of epidemically relevant adenovirus types is of interest not only for the country as a whole but also for individual areas, so as to establish, among other things, the frequency of their succession at different periods of observation. Thus, adenovirus types 2 and 6 have now proved relevant for the Ukraine, where types 3 and 7 were important in the previous five-year period. In the Far East, adenovirus serotype 3 was most prevalent over the period of ten years. Adenovirus

Table 1 The dynamics of succession of epidemically relevant serotypes of ARD adenoviruses isolated in the USSR, 1976–1986

Year	*Most relevant adenovirus serotypes*
1976	3, 7
1977	2, 3, 7
1978	2, 3, 7
1979	3, 2, 7
1980	3
1981	3, 6, 7
1982	3, 2
1983	2, 3
1984	2, 6, 1
1985	6, 1, 2, 3
1986	5, 6

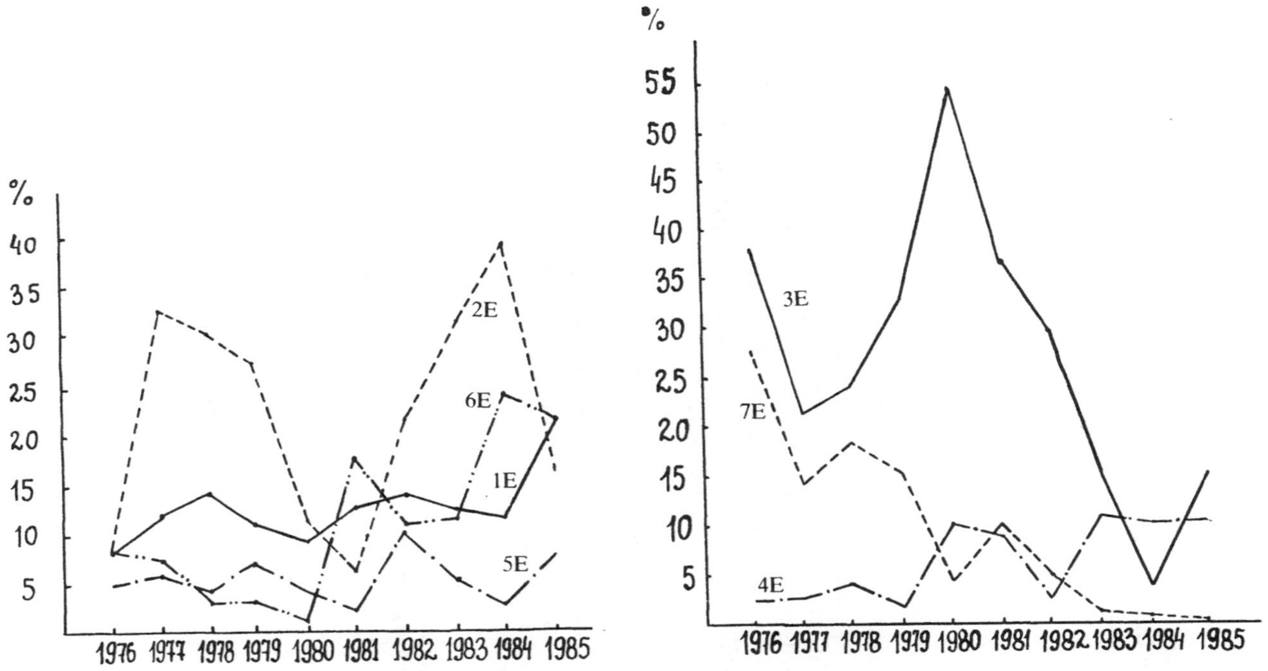

Figure 3 The dynamics of circulation of different types of adenoviruses in the USSR as a function of observation period. The abscissa is the years of observation. The ordinate is the percentage isolation of adenovirus types among the total number of adenovirus isolates. 2E, type 2; 6E, type 6; etc.

type 7 retained its relevance only in the north of the Russian Federation. Most areas of the USSR, with the exception of the Far East, and of the north of the Russian Federation, have thus seen the succession of latent types 2 and 6 as the epidemically relevant adenoviruses (Table 2).

A study of the age dynamics of incidence has shown that adenoviruses of different serotypes affect children and adults alike with almost the same frequency.

2.2 Picornaviruses

Not infrequently the development of acute respiratory disease is etiologically associated with picornaviruses (ECHO and type B Coxsackie). Several serotypes of ECHO (types 19 and 6) and Coxsackie B (type 3) viruses – and less commonly ECHO serotypes 13, 12, 6, 7, 20 and 11, and Coxsackie B serotypes 1, 5 and 6 – are crucial to the genesis of ARDs. The proportion of these serotypes varies with the period of observation. To illustrate, relevant types of ECHO viruses have recently been replaced with ECHO 6 virus succeeding ECHO 19 virus. As far as Coxsackie B viruses are concerned, serotype 3 remained relevant over the observaton period of ten years.

Picornavirus illnesses will recur in summer, as was manifest between 1976 and 1985 in most regions of the USSR except for the Volga area and the Urals. However, recent years have seen a shift in the seasonal recurrence of picornavirus infections. For example, in 1982 and in 1984,

Table 2 The dynamics of succession of epidemically relevant adenovirus serotypes in different climatic and geographical zones of the USSR as a function of observation period.

Region	*No. of strains studied*	*1976–1980*	*1981–1985*
Ukraine	977	3, 2, 7	2, 6
North-western Russian Federation	184	2, 3	2, 6, 1
Far East	108	3, 2	3, 1, 6
Central European part of Russian Federation	167	3, 7	2, 3, 6
Northern Russian Federation	60	3, 7	3, 7, 1
Urals, Volva, Region, Byelorussia, Western Siberia	196	3, 2, 7	2, 1
Total	1712	3, 2, 7	2, 6, 1

the greatest proportion of picornavirus infections occurred at a period of seasonal rise in morbidity.

Picornaviruses were more often isolated from children between age 3 and age 7.

2.3 Respiratory syncytial virus

RS virus is characterized by epidemic periods of increased incidence. Periods of maximal incidence vary with age groups, averaging six months (November–April) in children between ages 0 and 2, four months (December–March) in children between ages 3 and 6, and three months (December–February) in school children and adults. In this period, the incidence is 2.5–10 times above the ordinary one.

Intensive epidemic RS outbreaks of viral etiology are followed by 8 to 9-month periods of low viral activity, and, in contrast, the low attack rate by the agent in an RSV epidemic period promotes the occurrence, within the year, of two or three rises in incidence over two-month intervals.

The respiratory syncytial virus evokes ARD outbreaks annually. The highest incidence occurs between October and March. In the event of RS infection, the intensity of the epidemic process varies annually over a wide range, from low to high. If it is high, there are many ARD outbreaks in the community, the incidence and prevalence rates are high, clinically pronounced forms, not uncommonly with complications (otitides, etc.), are numerous, and asymptomatic infection forms are fewer.

In children, in particular young children, the RS virus frequently induces reinfections even within the same year. Their annual prevalence rate is dependent upon the intensity of the epidemic process and varies, on average, from 5 to 19% among institutionalized children. Some children contract RS infection repeatedly within the year.

The period when the causative agent can be isolated from the rhinopharynx of ARD patients lasts between three and 40 days. In 30% of cases the virus can be isolated within 20–40 days.

2.4 Parainfluenza viruses

In the USSR, the proportion of parainfluenza among ARDs averages 9.9 to 12.2%. Many years of study into the time course of the circulation of parainfluenza viruses (serotypes 1 and 2) have revealed that they produce ARD outbreaks every two to three years. There is a well-defined succession of these agents in the circulation dynamics, since the intensive prevalence of serotype 1 parainfluenza is accompanied by a modest prevalence of serotype 2 parainfluenza, and *vice versa*. It should

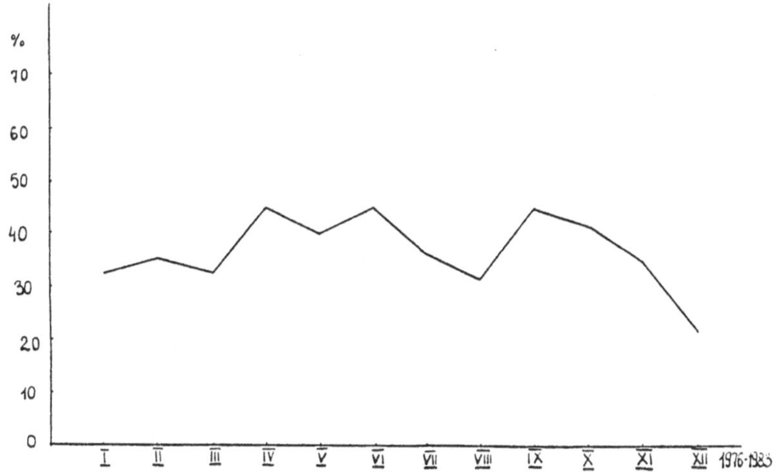

Figure 4 The dynamics of circulation of serotype 3 parainfluenza virus at different seasons (overall data from serological diagnosis between 1976 and 1983). The abscissa is the years of observation. The ordinate is the percentage of seroconversion.

be noted that the proportion of serotype 1 influenza virus was somewhat greater than that of serotype 2 influenza virus at all the seasons of the year (averaged data).

These agents were isolated mostly during the period of seasonal increases in incidence.

As far as serotype 3 parainfluenza virus is concerned, its prevalence rate was two times higher than that of other parainfluenza viruses. This virus was characterized by a bimodal prevalence with two peaks of incidence occurring in the period of seasonal rise and in the summer months (Figure 4). For instance, maximum rises in the incidence of this agent occurred in April, September and June (averaged data).

3. THE BASIC VARIABILITY PATTERN OF RESPIRATORY VIRUSES IN THE COURSE OF THEIR EPIDEMIC CIRCULATION

In nature, viruses evolve in accordance with their antigenicity and biological properties, giving rise to a multitude of strains and variants with different phenotypical and genotypical profiles. Genomic variants are known for a broad spectrum of RNA- and DNA-containing viruses, such as rotaviruses (Ranon and Holmes, 1985), blue tongue viruses (Oberst *et al.*, 1985), baculoviruses (Brown *et al.*, 1985), and herpes simplex (Sakaoka *et al.*, 1985).

The climatic, geographical and natural peculiarities of different areas have a great influence on the selection of strains with a specific genotypical characteristic. Geographical strains of arenoviruses and granulosis viruses, differing from their prototypes by a marked change in their genome structure, have been discovered (Gonzales *et al.*, 1985; Shelulhina *et al.*, 1979).

Geographic virus strains may differ from one another in biological properties and virulence. Studies into the virulence of pox viruses (Shelulhina *et al.*, 1979; Tantawi *et al.*, 1978) and togaviruses (Trent *et al.*, 1978) revealed a clearcut relationship between the properties of strains and their geographical origin.

However, there is little information about the variability pattern of respiratory viruses in the course of their epidemic circulation – a subject central to all our investigations.

3.1 Population analysis of circulating adenovirus strains by antigenicity and some biological properties

Analysis of natural populations of serotype 3 adenovirus in selected population samples, isolated in different areas of the USSR between 1976 and 1982, has revealed their antigenic heterogeneity (Table 3). A mere 34% of the viruses were similar to a type 3 reference strain (GB). On the other hand, 17% of the strains differed markedly from their prototype, interacting with a reference serum down to 1/16–1/32 of its homologous titre. A type 7 adenovirus population had up to 70% resemblance to the reference virus, and only 3% of the strains differed considerably from it. Apparently, the population of serotype 3 adenovirus strains circulating in the USSR consisted of two subtypes one of which, subtype 3, was similar to the reference virus, while the other, subtype 3a, was appreciably different from it. The heterogeneity of the population in nature is assumed to be maintained by the simultaneous occurrence in it of the two subtypes whose prevalence was dependent upon the particular observation period. Thus, for instance, in 1978 and in 1981, antigenically different variants made up 24 and 40%, respectively. It is to be noted that the prevalence rate of highly toxic adenovirus strains proved to be maximal (Table 4). The prevalence rate of highly toxic strains appears to have been dependent on the predominance in the population of subtype 3a adenoviruses. We have analyzed 30 isolates by use of restriction endonucleases (REs) Bam HI and Mind III and found three genome types Ad3au, Ad3ag and Ad3a10.

Table 3 Antigenic profile [a] of serotype 3 adenoviruses isolated in different years.

Year	*No. of strains studied*	*HT*		*1/2 HT*		*1/4 HT*		*1/8 HT*		*1/16 HT*		*1/32 HT*	
		Ab	*%*	*Ab*	*%*	*Ab*	*%*	*Ab*	*%*	*Ab*	*%*	*Ab*	*%*
1976	11	1	9.1	2	18.2	—	—	7	63.6	1	9.1	—	—
1977	17	1	5.9	3	17.6	8	47.1	5	29.4	—	—	—	—
1978	25	2	8.0	5	20.0	9	36.0	3	12.0	5	20.0	1	4.0
1979	9	—	—	4	44.4	—	—	4	44.4	1	11.2	—	—
1980	14	4	28.6	3	21.4	4	28.6	1	7.1	2	14.3	—	—
1981	15	1	6.7	4	26.7	4	26.7	—	—	4	26.7	2	13.3
1982	3	—	—	2	66.7	—	—	1	33.3	—	—	—	—
Total	94	9	9.6	23	24.5	25	26.6	21	22.3	13	13.8	3	3.2

[a]Described by the number of strains interacting with serum to the reference strain to the fraction of the homologous titre (HT) shown. Ab = No.

Table 4 Levels of cytotoxic activity in HeLa cell cultures of serotype 3 adenovirus strains isolated in the USSR.

Observation period	*No. of strains studied*	*Slightly toxic*	*Moderately toxic*	*Highly toxic*
1976	16	62.5	31.2	6.3
1977	43	48.8	39.5	11.7
1978	49	43.0	28.5	28.5
1979	30	70.0	13.3	16.6
1980	12	83.4	8.3	8.3
1981	21	81.0	9.5	9.5
Total	171	58.5	25.1	16.4

Values are percentages of strains with the indicated level of cytotoxic activity.

Japanese researchers have obtained similar evidence to show that, as opposed to serotype 3, the subtype 3a adenovirus has tended to induce clinically manifest forms of infection (Wadell, 1984).

The structure of the adenovirus (serotype 3) population varied as much with the period of observation as it did with the particular climatic and geographical area. There was a rather high rate of prevalence for highly to moderately toxic strains in some areas of the USSR, their proportion being greatest in the Urals and in the Far East.

As regards serotype 7 adenovirus, it has now lost its relevance in USSR territory – this correlates with the antigenic homogeneity of the naturally occurring serotype 7 adenovirus population.

3.2 The biological profile of a naturally occurring RSV population, its variability with the rate of development of the epidemic process, and the effect of individual immunogenetic factors of the body on the prevalence of strains with specific genotypical and phenotypical characters

A natural RSV population is reasonably polymorphic in biological characters such as reproduction in tissue cultures at optimal and supraoptimal incubation temperatures, degree of sensitivity to antibodies, and active interference with the vesicular stomatitis virus (VSV) in tissue cultures.

Highly virulent strains showed, as a rule, high reproduction rates in tissue cultures (infectivity regression coefficient; IRC), poor sensitivity to antibodies, and low interfering activity. RSV composition would vary with the period of observation from a generation of highly virulent strains to that of low-virulent variants. The most virulent strain populations circulated in the 1976–1977 season, and the least virulent in the 1979–1980 season. In different years, the population biological profile correlated with the high to low rate of development of the epidemic process in a community of institutionalized children. For example, in the 1976–1977 season, a population of high-yielding, low-avid, and non-interfering RSV strains induced the most intensive ARD outbreaks among young children. Thus, over the years of observation, the dynamics of prevalence of viral populations changed from high-yielding strains in the period 1976–1977 (60–80%) to moderately-yielding variants between 1977 and 1978 (58–88%), and finally to low-yielding varieties in the period 1979–1980.

It should be noted that the individual immunogenetic factors of the organism had a considerable influence on the prevalence rate of strains with a certain phenotypical profile.

The biological profile of the RSV viral population varied continuously with time with respect to the IRC_{39} (at 39°C), which is in line with the continuity of the epidemic process and its variability from high to low intensity of manifestation. Succession of RSV populations by the IRC_{39} characters reached maximum values five to six months after the termination of RS outbreaks in a community of institutionalized children. This would normally coincide with a summer period of observation. A population of summer strains showed low infective potential. Subsequently, with the onset of fall, the strains tended to change so as to become capable of reproducing at elevated incubation temperatures.

In terms of the intensity manifestation, then the IRC_{39} genetic character changed in the definite direction of $+++ \rightarrow ++ \rightarrow + \rightarrow$. During the next epidemic season, there was a change in reproducing performance of the strains at a variable reversion level in the opposite direction: $+ \rightarrow ++ \rightarrow +++$. This is testimony to the recurrence of the natural variability of the respiratory syncytial virus and to particular population waves being responsible for the phenotypical structure of naturally occurring viral populations and the rapid succession of their generations.

3.3 The polymorphism of naturally occurring populations of serotype 3 parainfluenza viruses

There are isolated reports about a natural population of parainfluenza viruses (serotypes 1 and 2) being heterogeneous in their antigenic and biological profiles (Rybinskaya, 1975, 1976). However, no information is available concerning the pattern of the natural population variability of serotype 3 parainfluenza viruses.

Comparative analysis of antigenic and biological properties of current isolates of the serotype 3 parainfluenza virus, and those of the C-243/57 reference strain, have revealed marked distinctions between them.

Of the research strains isolated between 1977 and 1978, some were differentiated by particular biological and molecular and biological characteristics. For instance, unlike the 1978 strains, variants isolated in 1977 elicited inflammation of the lungs in newborn hamsters, a more pronounced local inflammatory reaction, more extensive leukocyte proliferation, and a productive type of infection in the alveolar epithelium of the lungs in newborn hamsters, as shown by electron microscopy. By contrast, strains isolated in 1978 induced an abortive type of infection in this system. Strains isolated in 1977 multiplied better in cellular tissue cultures and also possessed higher hemolytic activity.

It was possible to differentiate strains not only by biological but also by antigenic traits. Strains isolated in 1978 were similar to prototype (C-243/57), whereas the isolates of 1977 differed from it markedly (8 to 16-fold difference).

It should be specially emphasized that the 1977 population was epidemically more active (higher prevalence rate and higher reinfection rate) among institutionalized children as against the 1978 population.

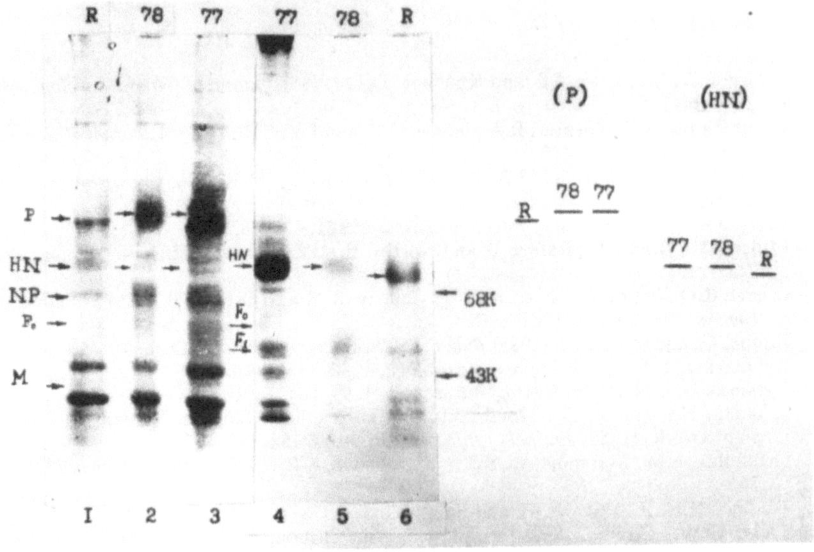

Figure 5 Electrophoresis in a 7.5% polyacrylamide gel of fractions of nucleocapsid proteins (1, 2, 3) and surface glycoproteins (4, 5, 6) after treatment of serotype 3 (PG3) human parainfluenza viruses with a 2% nonionic detergent, NP-40; 1, 6, prototype C-243/57 strain (R); 2, 5, PG3/1364/78 isolate (78); 3, 4, PG3/645/77 isolate (77). The arrows indicate the positions of markers for molecular weight (bovine serum albumin 68K and ovalbumin 43K).

The results obtained indicate that the natural populations of parainfluenza viruses differ in virulence.

We sought to compare strains differing in virulence by an electrophoretogram of their proteins using the prototype C-243/57 strain, a 1977 isolate (PG3/645/77), and a 1978 isolate (PG3/1364/78). Polyacrylamide gel electrophoresis revealed that the 1977 and 1978 isolates differed from one another only slightly in the profiles of their virion proteins. Yet the two isolates studied showed lower electropheretic mobility of polymerase and hemagglutinin–neuraminidase proteins as compared to the prototype strain. These distinctions were most clearly manifest in the electrophoresis of fractions of the nucleocapsid proteins and surface glycoproteins of the above viruses (Figure 5).

In summary, analysis of the molecular and biological properties is essential for elucidating the structure of a naturally occurring population and the degree of its variability in the course of epidemic virus circulation.

REFERENCES

Brown, S.E., Maruniak, J.E. and Knudson, D.L. (1985) *Journal of General Virology*, **66**, 2431–2441.

Crook Norman, E., Spenser, R.A., Payne, C.C. and Leisy, D.J. (1985) *Journal of General Virology*, **66**, 2423–2430.

Gonzales, J.P., McCormick, J.B., Peters, C.J. and Killey, M.P. (1985) *Virus Research*, **3**, Suppl., 87.

Larry, J.A. (1985) *Journal of Infectious Diseases*, **151**, 626–633.

Mufson, A., Orwell, C., Rafner, B. and Norrby, E. (1985) *Journal of General Virology*, **66**, 2111-2124.

Oberst, R.D., Souri, K.N.E., Stott, J.L., Chuang, R.Y. and Osburn, B.I. (1985) *Journal of General Virology*, **66**, 1901–1909.

Ranon, G. and Holmes, I. (1985) *Bulletin of the Society of Pathol. Exot.*, **78**, 456–466.

Rybinskaya, L.N. (1975) *Voprosy Virusologii*, **1**, 30–34 (In Russian).

Rybinskaya, L.N. (1976) *Voprosy Virusologii*, **1**, 99–104 (In Russian).

Sakaoka, H., Aomori, T., Honda, O., Saheki, Y., Isida Shigeru, Yamanishi, S. and Fujinaga, K. (1985) *Journal of Infectious Diseases*, **152**, 190–197.

Shelulhina, E.M., Marennikova, S.S. and Froltsova, A.E. (1979) *Acta Virologica*, **23**, 360–366.

Tantawi, H.H., El-Dahaby, H. and Fahmy, L.S. (1978) *Acta Virologica*, **22**, 456–457.

Trent, D.W., Clewely, J.P., France, J.K. and Bishop, D.H.L. (1978) *International Virology*, **4**; *Abstracts of the 4th International Congress on Virology*, The Hague: Wageningen.

Wadell, G. (1984) *Current Topics in Microbiology and Immunology*, **110**, 191–220.

Sov. Med. Rev. E. Virology Reviews, Vol. 3. 1990, pp. 79–95
Photocopying permitted by license only

PRECURSORS OF INFLUENZA EPIDEMICS AND THEIR PRACTICAL APPLICATIONS

G.S. SKRIPCHENKO

I.I. Mechnikov Research Institute of Virology and Epidemiology, Odessa

and

A.N. SLEPUSHKIN

D.I. Ivanovskii Research Institute of Virology, USSR Academy of Medical Sciences, Moscow

Abstract

This paper summarizes the research into precursors of influenza epidemics conducted by the authors over many years. The findings are compared with those of other Soviet and foreign researchers. The emphasis is on the purpose-oriented identification of such precursors for the etiological prediction of influenza outbreaks in particular populated settlements, areas, or the country as a whole. The theoretical and practical aspects of established epidemiological patterns of influenza outbreaks are discussed.

Key words: precursors; community immunity; seroconversion

1. INTRODUCTION

Predicting influenza epidemics is still a difficult task. Numerous factors, however, have now accumulated which allow scientific approaches to the prediction in advance of possible influenza outbreaks for a forthcoming epidemic season. In our view the most informative of these possible approaches is based on identification of the precursors of such epidemics.

A purpose-oriented identification of this kind only becomes possible because of the outcome of lengthy research into the circulation of the influenza virus and its variability during non-epidemic periods. Studies conducted in different climatic and geographical zones of the Soviet Union in 1962–1985 have produced data comparable with that available in the literature and shows that the emergence of such precursors is conditional on the immunological and biological activation of an agent of disease during the latent period of each epidemic. By determining when this period begins and the serotype of the future epidemic agent through the identification of precursors, it will become possible etiologically to forecast an epidemic several months before its outbreak.

During the initial stages of research, systematic virological and serological examinations of more or less healthy people, as well as of outpatients and their contacts, revealed that in foci of sporadic influenza most of the people surrounding a sick person are infected and suffer the disease in subclinical form (Skripchenko, 1963, 1966, 1969). Such persons, in turn, infect people around them. During non-epidemic periods, however, new cases of illness in such foci usually revealed themselves after very long intervals following their contacts with a sick person. As a rule, manifestations of illness occurred after exposure to adverse temperatures (overheating followed by overcooling). The intensity of the formation of foci of symptom-free influenza, caused by the pathogen of a developing epidemic, was found on average to increase 3–4 months prior to the increase of morbidity. The period of the agent's increasing circulation and the formation of an epidemic population was characterized by a certain consistency in its manifestations, which allowed them to be regarded as precursors of an epidemic (Skripchenko, 1966). Depending on the method of identification, these precursors may be conventionally broken down into three types: immunological, virological and epidemiological.

2. IMMUNOLOGICAL PRECURSORS

Among the immunological precursors of epidemics the data are summarized in Skripchenko, 1970; Skripchenko *et al.*, 1978) one needs to mention, first of all, a predominant seroconversion among practically healthy people to the virus whose subsequent variant would eventually cause a rise in morbidity of epidemic proportions (Figure 1). Thus, in the period preceding the influenza B epidemic outbreak in January 1963, an increase in antibody titre to this virus type was observed in paired blood sera of practically healthy people four times more frequently ($p < 0.05$) than to the influenza A(H2N2) virus. Prior to the influenza A(H2N2) epidemic, which took place in Odessa in 1965, a growing anti-hemagglutinin titre to influenza A(H2N2) virus in practically healthy people was registered. The frequency of a four-fold increase or more in antibody titre to A(H2N2) virus was 8.45 times greater ($p < 0.001$), compared to influenza B virus. Yet, in the period between epidemics

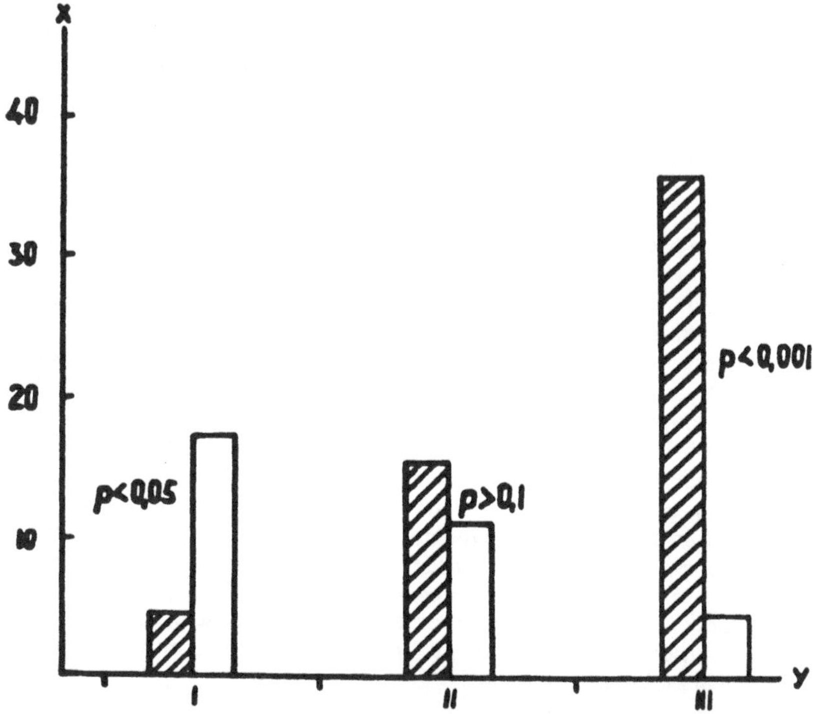

Figure 1 Peculiarities of seroconversion to A(H2N2) influenza (hatched columns) and to virus B influenza (white columns) in practically healthy people within a permanent workers' collective over different periods of time.
Periods on the *abscissa*: I, prior to influenza B epidemic; II, between the influenza B and A(H2N2) epidemics; IIi, prior to influenza A(H2N2) epidemic. *Ordinate*: Frequency of diagnostic seroconversion, (%).

there was no difference in the frequency of seroconversions to these viruses ($p > 0.1$).

An increase in the number of seroconversions among practically healthy people during pre-epidemic periods, pointed to the development of influenza in a subclinical form and resulted in a high titre of anti-hemagglutinins to these viruses in the human population.

Similar instances of an increasing circulation of influenza viruses were detected by random surveys of the general public, conducted in Odessa and Moscow during the intensive circulation of pandemic variants of the influenza A(H3N2) agent in 1969 and that of A(H1N1) virus in 1977.

In 1968, an increase in the anti-hemagglutinin level by 2.21 times ($p < 0.001$) in human blood serum to A(H2N2) virus was registered in Odessa four months prior to the Hong Kong influenza morbidity of January 1969 (Figure 2). At the same time, viruses similar in their antigen structure to the A/Hong Kong/1/68 strain were isolated from poly-clinical patients one month prior to the outbreak. During the development of a second epidemic of the same cycle next year, an increase by 2.15 times in the mean hemagglutinin titre ($p < 0.001$) was recorded two months prior to the rise in morbidity (Knyazeva *et al.*, 1975).

Researchers from the Institute of Virology, USSR Academy of Medical Sciences, reported the emergence of antibodies to the influenza A/USSR/90/77 (H1N1) virus in the titre reaching the ratio of 1 : 20 and higher in 25.9% of 18 to 20-year-old examinees (Figure 3) one month prior to the influenza A(H1N1) outbreak in 1977 (Slepushkin *et al.*, 1979).

It should be mentioned that after a rise in the anti-hemagglutinin levels, which had preceded the influenza outbreaks, by the time the morbidity began to grow the population's immunity rate had again declined (Skripchenko *et al.*, 1978). Decline in the population's immunity, just before each epidemic, represent, in our opinion a factor which activates the virus. Alternation of mounting pressure of directive selection with periods of attenuation, apparently constitute the biological basis of such activation. An increase in the selection pressure takes place in parallel with the population's growing immunity as a result of a "hidden" intensification of virus circulation, which is then followed by periods of its rapid attenuation, brought about by the organism's diminished resistance because of unfavourable climatic factors. Such situations create the preconditions for selecting epidemiologically active virus strains, since it is known that alternations in selection pressure intensities represent a factor which accelerates the microevolution of any biological species. (Schmalhausen, 1969; Mayer, 1974).

The correctness of these general principles of population genetics with respect to influenza virus was confirmed experimentally. Signs of agent

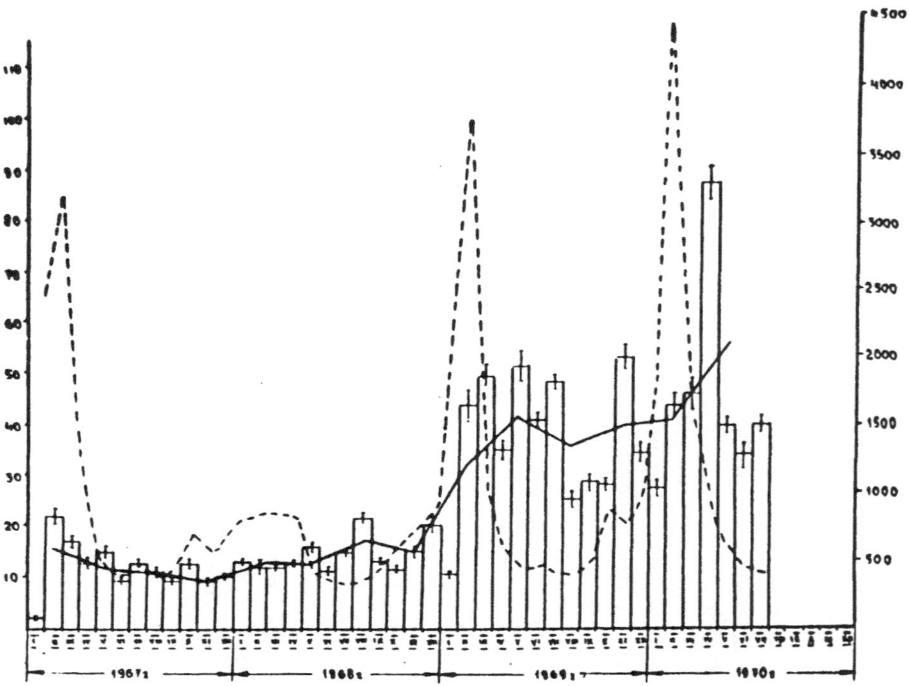

Figure 2 Dynamic changes in the documented incidence of influenza and acute respiratory diseases in comparison with indices of humoral immunity of the influenza virus A(H2N2) in Odessa. Average daily morbidity (broken lines) and average quarterly anti-hemagglutinin level in the population (solid lines); average monthly hemagglutinin level (columns); errors of mean anti-hemagglutinin titre (lines at the column tops) and serological type of epidemic (arrows at tops of morbidity peaks). *Ordinate*; reverse anti-hemagglutinin mean (left) and incidence of influenza and acute respiratory diseases (right).

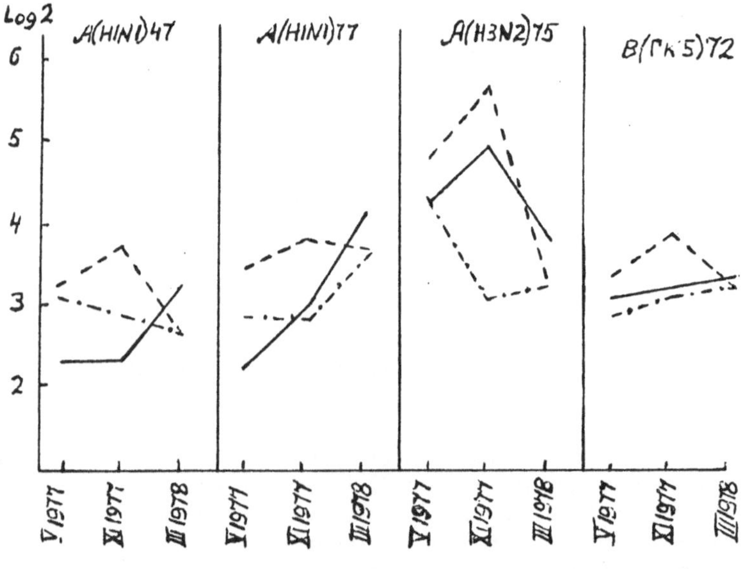

Figure 3 Dynamic changes of anti-hemagglutinin to virus A(H1N1), A(H3N2) and B influenza among Muscovites of different age in the period from May–November 1977 to March 1978. *Ordinate*: geometrical mean of antibody titre. Age brackets: 15–20 years of age (solid line); 21–35 years (broken line); 50 years and over (dot-and-dash line).

activation and continuous circulation were reproduced in variability stimulation tests by successively alternating pressures ranging from a highly intensive selection in the immune system to its attenuation during virus accumulation in conventional embryos (Skripchenko *et al.*, 1983).

The emergence of signs of immunological and biological activation, followed by continuous circulation of influenza virus, as well as stimulation tests for agent variability, prove that an influenza epidemic has self regulatory mechanisms which, according to Belyakov (1983), are intrinsic to all parasitic systems.

Immunological signs of virus activation and increased circulation prior to an approaching epidemic were detected even earlier by examining placental immunoglobulins. Thus, antibodies, in titres of 1 : 40 and higher, to the A/Victoria/35/72(H3N2) virus which came to dominate the etiology of influenza after 1972, were initially detected as early as 1970, i.e. two years before the virus was isolated. Antibodies to the A/Port Chalmers/1/73 (H3N2) virus which caused the 1974–1975 epidemic, began to be detected in the 1 : 40 titre back in 1972, again – as was the case with the "Victorian" strains – two years before these viruses came to dominate the epidemic process. Considerable changes in the antibody level to the "old" A/FM/1/47 strain were detected two years prior to the epidemic caused by the A(H1N1) serotype. By 1975 the antibody level had increased 2.3 times up against the 1969 figure ($p <$ 0.001). At the same time there was no rise in antibody level to the virus with a zero hemagglutinin type. It should be noted that these data were reported before the epidemic growth of incidence, caused by the A(H1N1) influenza virus (Skripchenko *et al.*, 1977).

3. VIROLOGICAL PRECURSORS

The next group of precursors which characterizes the specific development of an epidemic influenza virus population, has been associated with progressive domination of the agent of the incipient epidemic in the etiology of recorded diseases.

During the non-epidemic seasons (July–September 1962 and March 1962–November 1964) the participation of A(H2N2) and B viruses in the etiology of diseases was roughly the same ($p > 0.1$) (Figure 4). On the eve of the epidemic the number of pronounced cases of influenza caused by the A(H2N2) virus, exceeded that of influenza B cases by 16 times ($p < 0.001$). Prior to the influenza B epidemic it was possible to detect cases caused by this virus only ($p < 0.001$) (Skripchenko *et al.*, 1978).

It should be mentioned that increased circulation of influenza A virus and cases of the disease occurred at a mean anti-hemagglutinin titre of about 1 : 10. The further growth of the mean titre up to 1 : 20 and higher preceded a massive rise in influenza A incidence. In the case of influenza B, the corresponding figures are titres of 1 : 20 and 1 : 40. These data allow the claim that there exist conventional "minimal" values of population immunity strength, beneath which increased influenza virus circulation is possible, as well as "critical" values which may be used for assessing when the epidemic is moving into the stage of a clinically recorded rise in morbidity (Skripchenko, 1970; Skripchenko *et al.*, 1978).

The possibility of using this influenza precursor for predicting epidemics for the coming season was confirmed by Slepushkin, Fedorova and co-workers (Slepushkin and Fedorova, 1982; Slepushkin *et al.*, 1983a and b, 1984; Fedorova *et al.*, 1984). From 1978 to 1984, sera of some 500 people who lived in Moscow and 10–12 other Soviet cities situated in different geographical zones, was H1 tested each May or June in the laboratory of the Epidemiology Institute of Virology, USSR Academy of Medical Sciences, and of the USSR Regional Influenza Research Centre (Figure 5). Blood sera from five groups consisting of 50 healthy people and those who had had infectious diseases were collected

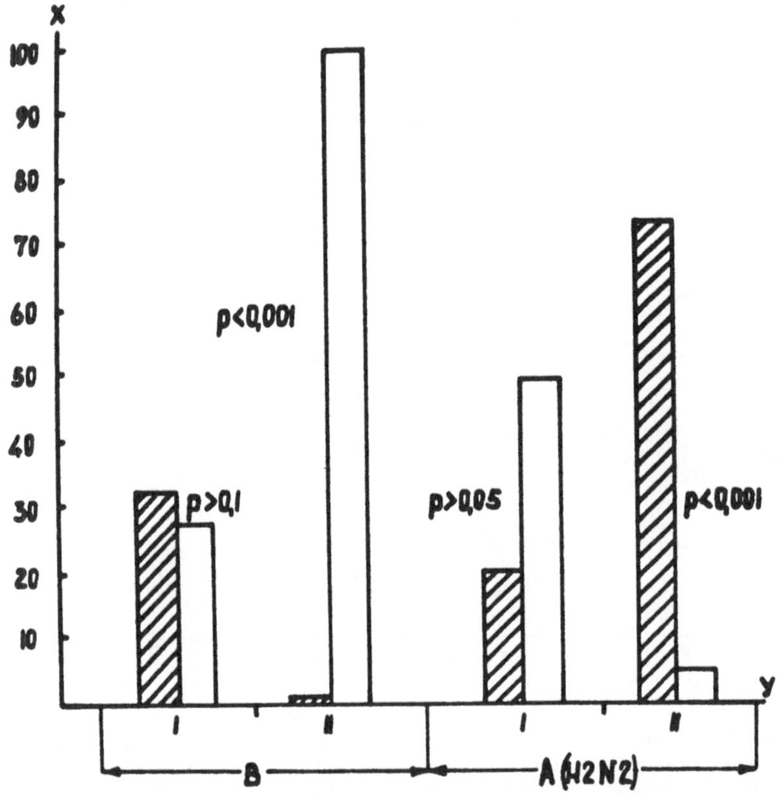

Figure 4 Peculiarities of seroconversion to viruses A(H2N2) and B in people who suffered from influenza during different stages of the epidemic. *Abscissa*: frequency of diagnostic seroconversion (%);*Ordinate*: B, prior to influenza B epidemic; A(H2N2), prior to A(H2N2) epidemic; I, inter-epidemic; II, pre-epidemic.

twice a year (in autumn and spring). The five groups consisted of children under 14 years of age, adults aged 15–25,26–40, 41–54 and those over 55 years old. In early summer the sera were processed and assayed simultaneously against antigens of influenza A and B viruses. To standardize the results a reference serum was used for the reaction (at the facilities of the USSR Regional Influenza Research Centre).

By analysing the antibody dynamics over a preceding season as well as the results of assaying the autumn serum prior to an epidemic season, it became possible, on the basis of antibody rises in the sera to the "critical" level, to predict influenza A(H2N2) epidemics in the 1979–1980, 1982–1983,1984–1985 seasons, as well as the influenza B epidemic

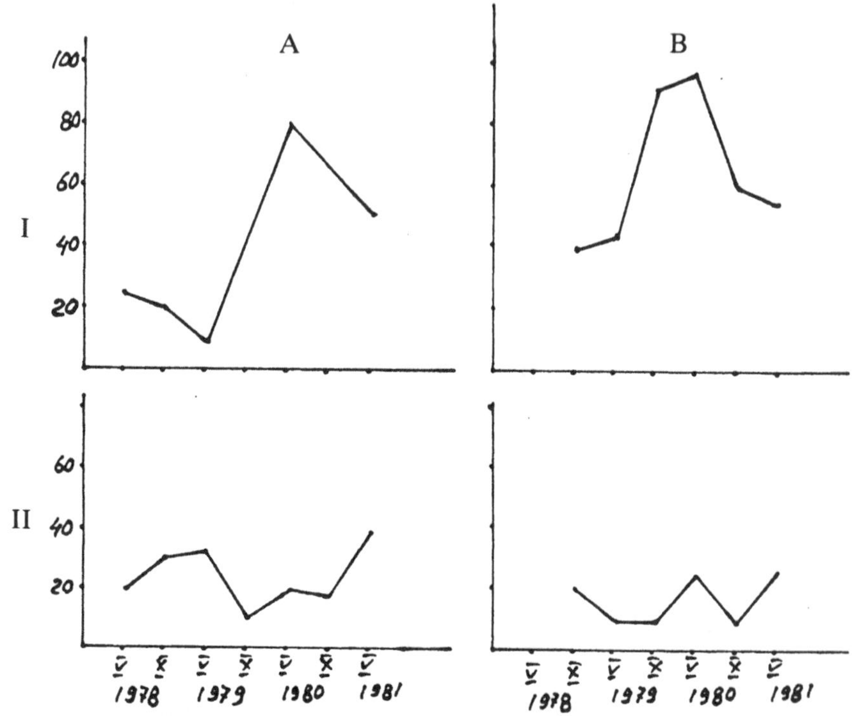

Figure 5 Dynamic changes of antibodies to influenza A and B viruses in 1978–1981. *Ordinate*: persons with anti-hemagglutinin titre 1 : 20 and higher (%). I, A/Texas/1/77/H3; II, b/Hong Kong/5/72; a, all examinees; b, examinees aged 15 to 20.

in the 1980–1981 and 1983–1984 seasons two to four months prior to their outbreak.

The ability of the circulating agent to stimulate the formation of antibodies and seroconversion to a higher level at each subsequent stage of the epidemic, may be considered the most important immunological precursor characterizing influenza virus activation and its antigenic novelty.

In particular, during the pre-epidemic season the titre of anti-hemagglutinins in people who had suffered influenza B was 4.62 times higher ($p < 0.01$) than in persons who had the disease in the period between epidemics. In case of A(H2N2) influenza, the intensity of antibody formation in response to infection during the pre-epidemic season was 1.59 times higher ($p < 0.01$) while during the epidemic it was 1.96 times higher ($p < 0.01$) than in the period between epidemics (Skripchenko, 1969; Skripchenko *et al.*, 1978).

The difference in the ratios between seroconversion to the A(H2N2) influenza virus during the inter- and pre-epidemic seasons was especially marked in comparison with the 1968–1969 epidemic periods when a change in the pandemic variation of the agent took place (see Table 1). Thus, as a result of hemagglutination-inhibition (HI) tests of paired sera conducted during inter- and pre-epidemic periods it was found that a

Table 1 Growth ratio of anti-hemagglutinin titre to A(H2N2) virus in persons serologically shown to be ill during different stages of the influenza A(H3N2) epidemic.

Parameter	*Pre- and inter-epidemic (May–December 1968, 1969, 1970)*	*Epidemic (January–March, 1969, 1970)*	*Statistically significant differences*
Number of HI tested paired sera	389	1438	
Frequency of diagnostical antibody rises seroconversions) found:			
absolute value	39	500	
percentage	12.02 ± 1.52*	34.77 ± 1.25	
Distribution of diagnostically found rises of antibody titres by ratio:			
4: absolute value	31	193	4-fold increase $T=5.99$ $p<0.001$
percentage	79.49 ± 6.47	38.60 ± 2.18	
8: absolute value	4	140	
percentage	10.25 ± 4.85	28.00 ± 2.01	
16: absolute value	2	88	
percentage	5.12 ± 3.53	17.60 ± 1.70	
32: absolute value	2	51	
percentage	5.12 ± 3.53	10.20 ± 1.95	
64: absolute value	—	28	
percentage	—	5.60 ± 1.03	
total 8-fold and over:			
absolute value	8	307	8-fold increase and over: $T=5.99$ $p<0.001$
percentage	20.51 ± 6.47	61.40 ± 2.18	

* $\text{percentage} \pm m = \sqrt{\frac{P(100-P)}{n}}$.

fourfold seroconversion predominated (79.49%), while during epidemics influenza patients had an eightfold and even more pronounced seroconversion (61.4%) – in both cases $p < 0.001$ (Skripchenko *et al.*, 1978).

4. EPIDEMIOLOGICAL PRECURSORS

It has become possible to identify the shortening of intervals between successive illnesses in influenza foci as an early epidemiological indication of the agent's biological activation. As already mentioned, in each influenza focus the virus disperses through contacts between the carrier and other people at home, at work, or wherever. During an inter-epidemic period, however, the majority of people who contract the disease show no specific symptoms, while the serum antibody level determined 2–3 weeks after contact with a sick person, will have kept rising and in some cases the influenza virus was released by the same people. Subsequent pronounced illnesses in the foci appeared relatively seldom over intervals lasting 2–3 weeks after contact with a virus carrier. By the end of the pre-epidemic period these intervals shortened to 1–3 days (Skripchenko, 1966, 1969). The shortening of the intervals between consecutive illnesses in foci and the increasing number of such foci in the course of an influenza epidemic could be documented at a stage much later than for the aforementioned precursors (2–3 weeks prior to epidemic growth in morbidity).

5. DISCUSSION.

Identification of precursors can be accepted as a major approach to etiological prediction regarding the probability of an influenza epidemic during a coming epidemic season. Such prediction should be based on an all-round epidemiological surveillance of morbidity. The importance of such surveillance is best illustrated by the following. The 1965 influenza A(H2N2) epidemic was correctly predicted in 1964, a success scored even before the principal propositions concerning epidemic precursors had been postulated (Zhdanov *et al.*, 1964 a, b), and was based on a comparison of the general epidemiological patterns of influenza, the results of studying the etiology of acute respiratory diseases, and the population's humoral immunity. Identification of the early heralds of influenza essentially boils down to the collection of data on changes in its etiology and in the level of community immunity during non-epidemic periods. Thus, systematic studies of influenza B and A, which began in 1962 and 1964 (Skripchenko, 1963, 1966), respectively, have been vigorously pursued ever since (Belyakov, 1986; Skripchenko, 1983). The results of further research into regulated patterns of epidemics allowed the identification of precursors for exogenous and endogenous influenza A and B epidemics in different climatic and geographical zones of the Soviet Union (Skripchenko, 1969; Skripchenko *et al.*, 1978; Slepushkin *et al.*, 1979; Zakstelskaya, 1969; Belyakov, 1976, 1984).

Universal precursors, which have been identified as a result of prolonged immunological, virological and epidemiological studies into the emergence of an epidemic population of the agent during epidemic-free periods are the following signs of increasing influenza virus circulation:

1. a growing number of sub-clinical cases during the development of an epidemic, accompanied by a rise of the population's levels of antibody against that serological type of the influenza virus whose later variant would subsequently bring about an epidemic rise in morbidity;
2. the growing domination of the agent of a developing epidemic in the etiology of influenza illness;
3. a growth of the community immunity to this virus from "baseline" to "critical" levels;
4. an increase in the ratios of seroconversions and antibody level among people who have recovered from this type of influenza at each sequel of the developing epidemic;
5. the shortening of intervals between repeat cases in the foci and a growing number of multiple foci.

Such purposeful identification of precursors has made possible the correct etiological prediction of influenza outbreaks, not only with respect to individual populated areas but even for the country as a whole (Slepushkin and Fedorova, 1982, Slepushkin *et al.*, 1982; Zakstelskaya, 1969; Ivannikov, 1971). This provided a basis for the practical utilization of such precursors within the system of epidemiological influenza surveillance (Skripchenko and Polyakov, 1977; Karpukhin, 1984).

It should be pointed out that our data on epidemic precursors and the coincidence of research findings conducted in different regions of the Soviet Union (Slepushkin *et al.*, 1979; Belyakov, 1976; Shadrin *et al.*, 1979; Denisov *et al.*, 1982) and elsewhere (Aymard, 1976; Glezen, 1982; Glezen *et al.*, 1982; Hannoun *et al.*, 1977; Hole-Simpson, 1979; Influenza Surveillance, 1977), provide evidence of the existence of a latent period for each endogenous or exogenous epidemic and the possibility of predicting the onset of an epidemic in the country where such influenza surveillance is conducted. While analysing data on epidemiological surveillance of influenza one has to bear in mind that the latent period and the period of a documented epidemic represent individual phases of an epidemic process with its own characteristic and specific manifestations. The first phase, consisting mainly of mild and subclinical forms, usually goes unregistered, and the second phase is manifested in clinically pronounced cases (Skripchenko, 1969; Skripchenko *et al.*, 1978).

Therefore, guided by this theory and using these precursors, one may predict an influenza outbreak by analysing the epidemiological situation and conducting routine immune epidemiological and virological studies both for large populated areas and the country as a whole. It is quite possible that the suggested approach may prove useful for predicting influenza outbreaks in forthcoming seasons on a regional or even global basis.

One may assume that ecological adaptation of the agent of the disease to local restricting factors takes place during the latent phase. In the course of this adaptation genetic transformations, which determine its heterogeneity according to known genetic parameters, are likely to occur in the population, influenced by the environment and specifically directed selection. Isolation of viruses, their identification and study of changes which indicate that the epidemic population of the agent has stopped development, may add significantly to our knowledge of such epidemic precursors. Some new immunological precursors, associated with dynamic changes in cell-mediated immunity in response to the growing circulation of the influenza virus which undergoes modifications in the course of each epidemic, may also be found.

In view of the foregoing, one of the approaches to etiological prediction of influenza outbreaks for the coming season is the method of active virological screening of sick persons and in-depth examination of isolated viruses as suggested by Glezen and co-workers (Glezen, 1982; Glezen *et al.*, 1982). The authors believe that at least 100 washings per week have to be examined for the purpose – thanks to such screening they were able, in 1976–1978, by the end of the preceding epidemics of influenza A(H3N2), B and A(H3N2), correspondingly to isolate virus strains for the coming season [influenza B, A(H3N2) and A(H1N1)]. Such data, in fact, coincide with the results of our own research. It should be mentioned, however, that virological screenings on such a grand scale require considerable outlays for enlarging existing virological laboratories, purchasing chick embryos, costly equipment and materials for isolation of viruses on different cell culture lines. Besides, such an approach to etiological prediction is one-sided and not fully representative, since it is based on identifying one group of virological precursors only. At the same time, it is common knowledge that the isolation of viruses (even when atypical in terms of antigenic structure) in the inter- and pre-epidemic periods does not by itself mean that the future epidemic would be caused by these viruses (Kilbourne, 1978; Schild and Stuart-Harris, 1967). At the same time, the simple system of surveillance and analysis we suggest is based on an identification not exclusively virological, but rather of several mutually complementary immunological and epidemiological precursors of influenza outbreaks. It could be organized, in the main, through joint efforts of the existing epidemiological and virological services available to the USSR Influenza Control Centre.

To sum up, the established pattern of agent activation observed during the latent developmental phases of influenza A and B epidemics, has made it possible to identify the major types of precursors and to show that the establishment of the epidemiologically significant virus population takes a long time. Identification of epidemic precursors during this period by sanitary and epidemiological services would facilitate the timely execution of an entire complex of specific and non-specific preventive measures against influenza, covering large territories and in accordance with the type of the developing epidemic.

REFERENCES

Aymard M. (1976) In: *Antigenic variants of Influenza Virus neuraminiase and anti-neurominidase*. Prophylaxis and Treatment of Influenza, pp. 17–38 Leningrad All-Union Influenza Institute Ministry of Health USSR (In Russian).

Belyakov, V.D. (1976) (Epidemiological backgrounds of Influenza immunoprofilactic). In: Inactivated Influenza Vaccine, pp. 91–97. Leningrad (In Russian).

Belyakov, V.D. (1983) Problem of self regulation of parasitic systems and mechanisms of establishment of epidemic process, *Vestnick AMN SSSR*, **5**, 3–9 (In Russian).

Belyakov, V.D. (1984) (Interconnection between Influenza epidemiology and construction of influenza vaccine). *Zh. Mikrobiol. Epidemiol. Immuno-biol.*, **3**, 115–9 (In Russian).

Belyakov, V.D. (1986) Slipping microbs. *Med. Gaz.*, February 7, No. 12, 4561 (In Russian).

Denisov, G.M., Naikhin, A.N., Ostrovskaya, A.D. and Umanskaya, L.V. (1982) Collective Immunity in the Framework of Epidemiological Surveillance over Influenza. In: Etiology and Diagnostics of Influenza and Other Acute Respiratory Diseases, pp. 5–12. Leningrad All Union Influenza Institute (In Russian).

Fedorova, G.I., Kiseleva N.V., Chvatov, P.P., Vitkina, B.S. *et al.*, Slepushkin, A.N., and Yagodovskii, V.V. (1984) In: Immunity of population of some cities of Soviet Union to Influenza viruses before and after epidemic of influenza in 1983–1984. *kvartal 1984 g.* Progress report of the Regional Influenza Center for the 3rd quarter of 1984, pp. 15–17 Moscow. Printed by I.D. Ivanoosky Institute of Virology AMS USSR (In Russian)

Glezen, W.P. (1982) Serious Morbidity and Mortality associated with Influenza epidemics. *Epidemiol. Rev.*, **4**, 25–44.

Glezen, W.P., Paine, A.A., Snyder, D.N. and Downs, T.D. (1982) Mortality and Influenza. *J. Infect. Dis.*, **146**, 313–321.

Hannoun, C., Aymard, M. and Menares, J. (1977) Le systeme de surveillance de la grippe en France. *Rev. Epidemiol. sant. publ.* **25**, 447–457.

Hole-Simpson, R.E. (1979) Epidemic mechanisms of type A Influenza. *J. Hyg. Camb.*, **83** 11–26.

Influenza Surveillance 1972–1975 (1977) *J. Hyg.*, **78**, 223–233.

Ivannikov, Yu.G. (1971) Methods of forecast of Influenza epidemics. In: Problems of Influenza. A Scientific Review, pp. 91–101. Moscow Institute of Medical Information, Ministry of Health, USSR (In Russian).

Karpukhin, G.I., Ed. (1984) Methodological Recommendations for Studies into Collective Influenza Immunity in the Framework of Epidemiological Surveillance over This Infection in This Country, pp. 1–22. Leningrad All-Union Influenza Institutes Ministry of Health USSR. (In Russian).

Kilbourne, E.D. (1978) Influenza Viruses and Influenza, pp. 526–585. Moscow.

Knyazeva, N.I., Nemets, Z.A., Skripchenko, G.S. and Chernevskaya, L.V. (1975) Dynamics of antihaemagglutinins to the A(H2N2) Influenza virus in population of city Odessa in period of appearance of new pandemic influenza virus A/Hong Kong/68/H3N2 In: Kiev, Gromashevsky Institute of Epidemiology and Infect. Diseases, Viruses and Viral Diseases, Issue 3, 45–47. (In Russian).

Mayer, E. (1974) *Populiatsiya, vidy i evolyutsiya* (Population, Species and Evolution; *translated from English*). Moscow: Mir (In Russian).

Schild, G.C. and Stuart-Harris, C. (1967) Antibodies in Children's Sera for serologically different recently isolated strains of Influenza Viruses A2 and B. *Br. Med. J.*, **4**, 1, 82–85.

Schmalhausen, I.I. (1969) *Problemy darvinizma* (Problems of Darwinism). Leningrad (In Russian).

Shadrin, A.S., Luzyanina, T.Ya., Popova, T.L., Karpova, L.S., Gulyeeva, V.N., Cybalova A.M. *et al.* (1979) The epidemiological features of Influenza A and B in longitudinally followed collectives. Topical Problems of Immunology, Virology, Chemotherapy and Epidemiology, pp. 173–176. Leningrad, Meditcoina (In Russian).

Skripchenko, G.S. (1963) Influenza morbidity and dynamics of antibodies in sera of healthy people. In: Respiratory Viral Infections, pp. 121–122. Moscow. Printed by Ivanovsky Institute of Virology AMS USSR (In Russian).

Skripchenko, G.S. (1966) Studies of Influenza Virus Circulation during inter-epidemic period in an Industrial Setting. *Abstract of DSc Thesis*, Odessa (In Russian).

Skripchenko, G.S. (1969) The precursors for forecast of the growing Influenza epidemic situation during latent pre-epidemic period. *Zh. Mikrobiol., Epidemiol. Immunobiol.* **II**, 75–80 (In Russian).

Skripchenko, G.S. (1970) Some problems of influenza prevention by live influenza vaccine. In: Influenza and Viral Respiratory Infections, Issue 4, pp. 160–166. Kiev. Gromashevsky Institute of Epidemiology and Infectious Diseases. (In Russian).

Skripchenko, G.S. (1985) Precursors of Influenza Epidemics, their biological meaning and practical study. In: Etiology and Epidemiology of Influenza Today, pp. 112–118. Leningrad All-Union Institute for Influenza, Ministry of Health USSR. (In Russian).

Skripchenko, G.S. and Polyakov, E.M. (1977) Methods of Predicting Influenza Epidemics, pp. 1–17. Odessa Mechnikov Institute of Microbiology and Virology (In Russian).

Skripchenko, G.S., Polyakov, E.M. and Knyazeva, N.I. (1978) Immunological precursors of Influenza Epidemics. *Zh. Mikrobiol. Epidemiol. Immunobiol.*, **5**, 70–77 (In Russian).

Skripchenko, G.S., Vlasova, A.G. and Rybakova, T.M. (1983) The influence of the host organism of the phase changes in population of influenza virus. *Vestnik AMN USSR*, **5**, 28–34 (In Russian).

Skripchenko, G.S., Vlasova, A.G., Boshchenko, V.Yu. and Bezbrozh, I.M. (1977) Evaluation of epidemic activity and variation of Influenza virus through the dynamics of antihaemagglutinins in placentary γ-glutinins. *Vrachebnoie delo*, **11**, 147–51 (In Russian).

Slepushkin, A.N., Fedorova, G.I., Sosthina, I.I., Khayrullina, R.Ya., Korshunova, N.G., Kiseleva, L.V. *et al.*,(1982). In: Dynamics of antihaemagglutinins to A(H1N1), A(H3N2) and B Influenza, viruses in winter – spring period of 1981–82. Progress Report of the Regional Influenza Center for the 2nd quarter of 1982, pp. 15–17. Moscow I.D. Ivanovsky Institute of Virology, AMS USSR (In Russian).

Slepushkin, A.N., Fedorova, G.I., Yagodovskii, V.V., Vitkina, B.S., Petrova, A.S., Toniants, E.G. *et al.*, (1983) Dynamics of population immunity to influenza in some cities of USSR in epidemic season of 1982–83. In: Progress report of the Regional Influenza Center for the 2nd quarter of (1983) Dynamics of population immunity to Influenza in some cities of USSR in epidemic season of 1982-83. pp. 21–23. Moscow. I.D. Ivanovsky Institute of Virology AMS USSR. (In Russian).

Slepushkin, A.N., Fedorova, G.I., Popova, N.S. and Melnichenko, E.I. (1983) Immunity to Influenza and forecast of Influenza incidence for the coming epidemic season. *Vopr. Virusol.*, **3**, 290–292 (In Russian).

Slepushkin, A.N., Obrosova-Serova, N.P., Shchastnyi, E.I., Pugayeva, V.P., Chernetsov, Yu.V., Selivanov, Ya.M. *et al.* (1979) Immunostructure population of Moscow to the A(H3N2), A(H1N1) and B Influenza strains circulating in 1977. *Vopr. Virusol.*, **2**, 175–180 (In Russian).

Slepushkin, A.N., Obrosova-Serova, N.P., Fedorova, G.I., Selivanov, Ya.M., Kupriashina, L.M. and Khvoshchova, V.M. (1980) The Immunity of population to influenza as reflexion of the epidemic process. In: Influenza Immunology and Prophylaxis Through Vaccination, pp. 95–102. Leningrad All-Union Influenza Institute, Ministry of Health USSR (In Russian).

Slepushkin, A.N., Ritova, V.V., Feklisova, L.V., Fedorova, G.I., Obrosova-Serova, N.P., Safonova, E.S. *et al.* (1984) Results of a two-year study of humoral immunity to Influenza A and B viruses in children under the age of 14 years in Moscow and its suburbs. WHO Bulletin, **1**, 51–58.

Zakstelskaya, L.Ya. (1969) Prognosirovanie grippa (Forecast of Influenza) *Med. Gaz.*, February 25, No. 17 (In Russian).

Zhdanov, V.M., Orlova, N.N. and Slepushkin, A.N. (1964) Immunoprofilactica of Influenza. *Med. Gaz.*, No. 80, 2357 (In Russian).

Zhdanov, V.M., Zakstelskaya, L Ya., Nesterenko, M.K., Antonova, I.V., Selivanov, Ya.M. (1964b) Influenza and ARD morbidity in USSR during III quarter of 1964. In: Progress report of the USSR Regional Influenza Control Center for the 3rd quarter of 1964, p.5. Moscow ID. Ivanovsky Institute of Virology AMS USSR (In Russian).

Sov. Med. Rev. E. Virology Reviews, Vol. 3. 1990, pp. 97–116
Photocopying permitted by license only

THE EPIDEMIOLOGY OF INFLUENZA

Yu.G. IVANNIKOV

All-Union Research Institute of Influenza, USSR Ministry of Public Health, Leningrad

Abstract

While analysing the latest data on the epidemiology of influenza, the focus here is on the social aspects of influenza epidemics and their impact on the population's health as a whole (influenza morbidity, post-influenza complications, additional somatic morbidity and mortality) as well as on the related economic losses. Based on the latest advances in general epidemiology, this study substantiates the modern theory of the epidemic process during influenza outbreaks, analyses the peculiarities of the specific epidemiology of influenza, including the infection's transmission mechanism, the distinctive immunity of seperate populations, and pathogen variability in determining the epidemiological process. Special emphasis is placed on the specifics of recent influenza epidemics; trends in their changes are described, and forecasts are made for the future.

Key words: morbidity; mortality, transmission, community immunity

1. INTRODUCTION

In recent decades, the intensive development of international and national systems of epidemiological surveillance of influenza have contributed to the accumulation of a wealth of facts, the analysis of which has made it possible to gain a better insight into the epidemiological regularities of influenza. Advances in virology and the ability to isolate and study circulating virus strains resulted in the deciphering of the etiology of all past influenza epidemics, starting from 1931, the year when influenza A virus was discovered by Shope. However, a real understanding of the peculiarities of the epidemic spread of influenza among humans could only be based on a quantitative analysis of the data describing the dynamics and intensity of influenza epidemics. Such research was undertaken in the Soviet Union where the epidemiological surveillance of influenza is based on the daily and weekly registration of all cases of influenza and other febrile respiratory illnesses, with the population's breakdown by age, as well as on data from virological and serological diagnostics of acute respiratory infections, including systematic immunological studies. The theoretical analysis of the findings obtained has rested on the use of a new definition of the epidemic process which synthesized the classic epidemiological concepts with a modern understanding of the interactions between host and parasite populations (Ivannikov and Ismagulov, 1983). According to this new definition, the epidemic process is a complex self-regulating ecological system, mediated through natural (and social) factors of interaction between genetically and phenotypically diverse and dynamically changing host (man) and parasite populations. This interaction is realized through the averagely stable elementary mechanism of parasite transmission from its source to a susceptible organism. While elaborating a modern theory, the acceptance of this definition understandably requires consideration of the specific epidemiological problems of influenza, the impact of natural and social factors on the epidemic process in its historical aspect, and of the role and correlation of the basic factors determining changes in the epidemic process.

2. SOCIAL ASPECTS OF INFLUENZA

The importance of influenza, among all the illnesses besetting man, in terms of its social impact and tremendous damage to the national economy and the health of the population has long been recognized.

Influenza morbidity, combined with the incidence of influenza-like infections, exceeds the overall morbidity from all other infections. Not a single human illness could presently be compared with influenza in this respect. Thus, some 10 to 30% of the employed population's temporary incapacity is due to influenza and other acute respiratory illnesses. In some years, influenza and acute respiratory illnesses amounted to 40% of all adult illness, over 80% of all infectious pathologies and over 60% of children's morbidity. Thus we see that the main damage caused by influenza is to the health of the population. Influenza frequently takes a grave form and in some cases has a lethal outcome. Influenza patients often develop various complications, mainly of the respiratory organs, leading to a protracted course with long periods of disability. Moreover, it has now been proved without doubt that influenza gravely complicates the course of chronic diseases, affecting primarily the cardiovascular system and the lungs, adding to the population's morbidity and mortality. All preventive and anti-epidemic, as well as curative measures, involve considerable expenditure of manpower and material resources. Finally, the social security costs and loss of working time due to influenza incidence inflict huge economic losses.

Influenza morbidity depends on the virus type: A, B or C. Epidemic or non-epidemic morbidity are distinguished. Non-epidemic influenza is made up of sporadic cases and outbreaks in organized groups of children or, less frequently, of teenagers. Epidemic influenza may assume two forms, namely, epidemics in some individual cities or countries, and pandemics. Influenza C is characterized by sporadic incidence and also outbreaks among young children. The number of influenza C patients is not usually large and presently no quantitative estimates are available. By total number of cases, influenza A leads the field, while influenza B comes second. According to data collected over many years on the population's morbidity caused by acute respiratory illnesses, cases of influenza B amount to 4%, while the share of patients with influenza A is about 21%.

Continuous outbreaks of the disease in groups of young children or teenagers, which tend to become less frequent in summer time, are typical of influenza B. Country-wide epidemics of the disease which mainly affect children, occur once in two or three years. There has never been a single pandemic of influenza B affecting the entire world in the course of one year. The world-wide pattern of influenza B distribution is characterized by a gradual spread of the virus through countries and continents. The epidemic may take several years to develop. Influenza B may cause epidemics of a kind which would not spread to all the country's urban centres but would rather affect a few of them, sometimes only one or two towns.

Typically, an epidemic process caused by influenza A consists, first, of epidemics in cities and towns which then merge to form a country-wide epidemic, this, in turn, comes to be seen as part of the influenza A pandemic. Such pandemics, caused by shifts of the virus, spread as a rule throughout the globe in the course of one year or even less. In the between-epidemics seasons the incidence of influenza A is not high, and consists only of sporadic cases or outbreaks.

Table 1 provides an adequate idea of the population's morbidity during influenza A and B epidemics in Leningrad.

Generally speaking, even in the years of acute epidemics, influenza morbidity amounted to only 40% or less of the annual morbidity attributed to acute respiratory illnesses in the population. In conducting

Table 1 Characteristics of influenza A and B epidemics in Leningrad, 1953–1986.

		Percentage influenza morbidity				
Epidemic year	*Etiology*	*Age 0–2*	*Age 3–6*	*Age 7–14*	*Age 15 and over*	*All groups*
1953	A(H1N1)	27.0	—	—	5.4	5.7
1955	B	7.1	—	—	2.5	2.8
1956	A(H1N1)	28.7	—	—	10.9	12.6
1957	A(H2N2)	32.6	—	—	19.4	22.8
1959	A(H2N2),B	16.2	—	—	19.6	22.9
1961	A(H2N2)	9.6	—	—	8.8	8.5
1962	A(H2N2),B	26.5	—	—	13.9	18.7
1963	B	9.3	—	—	3.5	4.7
1965	A(H2N2)	30.2	—	—	13.9	16.7
1966	B	10.8	19.7	18.4	4.2	7.2
1967	A(H2N2)	16.8	14.4	11.0	8.9	10.1
1969	A(H3N2)	16.6	15.5	19.9	17.5	18.1
1970	A(H3N2),B	39.6	38.5	30.5	16.1	20.3
1971	A(H3N2)	22.1	10.9	7.9	3.5	4.7
1972	B	19.0	9.8	5.3	0	1.2
1973	A(H3N2)	34.2	25.0	12.0	15.7	16.2
1974	B	4.6	13.5	12.8	1.0	3.0
1975	A(H3N2)	15.3	22.2	15.0	6.5	6.3
1976	A(H3N2)	14.0	23.2	21.6	8.4	10.9
1977	B	13.8	28.2	22.8	5.3	8.0
1978	A(H1N1)	7.7	12.1	22.3	4.6	6.7
1978	A(H3N2)	8.9	11.5	4.7	0	0.8
1979	A(H1N1)	6.3	14.4	23.2	1.0	4.0
1979	A(H3N2)	11.5	16.5	10.1	3.7	5.0
1980	B	23.9	36.0	40.7	5.7	7.9
1981	A(H1N1)	6.0	17.2	14.5	0.2	7.7
1983	A(H3N2)	15.6	26.2	28.9	4.4	8.6
1984	B	33.8	43.8	43.8	5.2	11.2
1985	A(H3N2)	38.8	35.4	28.8	5.3	9.1
1986	B	14.9	24.3	32.7	3.9	8.6
1986	A(H1N1)	1.8	1.4	1.9	0.3	0.5

anti-epidemic measures, the major difficulty lies in overcoming the peak loads on primary care facilities as a result of the concentration of the total influenza cases over a brief period of time. As a rule, it takes only a few days for an outbreak to develop in a working collective, one month for an epidemic to seize a town, and 3 to 4 months to spread throughout a country. It should, however, be taken into account that the number of registered patients is unevenly distributed in the course of an epidemic. For example, an analysis of an incidence involving 28 million patients resident in 32 Soviet cities, revealed that, during an epidemic peak, some 54% of influenza cases were registered within three weeks on average. This means that 8% of the entire population (or about 2 million) contract the disease. Influenza morbidity in these cities varied within 2.4–3.8%. In other words, during an epidemic peak the total number of patients comes to almost 2 million people in urban communities alone (Ivannikov and Ismagulov, 1983).

The data on morbidity, based on patients presenting to primary care facilities, fall short of the real situation. Special random surveys have demonstrated that not all patients seek medical assistance. Thus, some 30% of adults and 20–25% of schoolchildren refrained from calling the doctor during influenza epidemics.

Influenza upsets the functioning of the organism's major systems. Such dysfunctions are mainly due to intoxication affecting metabolism, the blood, the cardiovascular and respiratory systems, the central nervous system and immunity. For this reason influenza aggravates the course of chronic illnesses and triggers numerous complications. Since it is the upper respiratory tract which is the portal of entry of the virus and the site of its intensive reproduction, illnesses of the respiratory organs — pneumonia, bronchitis and diseases of the ears, nose and throat — rank first among post-influenza complications (Zlydnikov and Smorodintsev, 1974). Cases of pneumonia have been registered with 6–16% of outpatients, mostly people over 50 years of age. The percentage of pneumonia is higher, up to 50%, with hospital patients. Some complications, up to 3%, end lethally. Compared to the lungs, complications affecting the ears, nose and throat, are less frequent — up to 3% of cases — and impairments of the nervous system are less frequent still. The latter, however, may be especially severe and assume the form of encephalitis, meningitis and polyneuritis (Martynov, 1970).

Among hospitalized influenza-stricken children, complications in the form of pneumonias have been observed in 50–60% of cases. Half of this group are infants under one year, while among outpatients similar complications affect 7–15%. Other acute complications (diseases of the ear, nose and throat, pyelitis, cystitis, etc.) amount to 5–6% of cases.

Influenza mortality, having reached its peak during the 1918–1919 pandemic (up to 1% in some countries) has been gradually declining ever since. This decrease may be explained by the fact that, due to growing urbanization and intensification of international links, the number of people who had once had contact with the influenza virus and retained residual immunity, has also increased. Improvements in living conditions, the discovery, first of sulfa drugs and then of antibiotics, and their application to the treatment of post-influenza complications, improvements in medical care, including methods of intensive therapy, have all produced a positive impact. Mostly, it is either young children or old people who died from influenza.

The consequences of an influenza infection do not always confine themselves to post-influenza complications; they may considerably aggravate different somatic diseases and even provoke, or represent one of the etiological components in the manifestation of previously hidden diseases. Therefore, a rise in the population's morbidity as a result of several somatic diseases has normally been observed either during influenza epidemics or immediately afterwards (depending on the nosological form of the disease in question). Thus, over the 1968–1971 period, Leningrad population's morbidity exceeded 6%, but this was not evenly distributed in terms of the nosological forms concerned: over 60% of the entire excess morbidity was due to cardiovascular diseases, over 20% to pulmonary diseases, the remainder being due to tonsillitis and laryngological diseases. It is interesting to note that if the number of acute illnesses (hypertension, bronchitis, acute pneumonia) increases during an influenza epidemic and reaches its maximum during the epidemic peak, then chronic illnesses reach their maximum additional morbidity beyond the epidemic peak after a lag of 20–40 days.

Influenza epidemics are accompanied by additional mortality as a result of influenza itself and pneumonia, as well as by the total increase in general mortality, and additional mortality due to a number of somatic diseases. The highest share of the additional mortality has been associated with cardiovascular diseases. Thus, vascular lesions of the brain may produce up to 40% of all additional mortality, with the cardiovascular diseases accounting for as much as 33%, and acute pneumonia for up to 15.6%. Malignant tumors, TB and some other causes also make notable contribution to the total additional mortality. An analysis of lethality, calcualted on the basis of additional mortality indices, shows that in this respect epidemics of influenza B and epidemics of mixed etiology are regarded as the most severe. Thus, considering the population's additional morbidity and mortality during epidemics, it goes without saying that influenza has been a major factor in determining the population's health as a whole. If the peak load falling on the primary care facilities during influenza epidemics is predictable and may be reflected in anti-epidemic contingency plans, the so-called epidemics of cardiovascular, pulmonary and other kinds of disease, which accompany influenza epidemics, have not so far been taken into account by the health authorities, although the resultant mortality from this latter group of diseases is probably even more damaging than the influenza epidemics *per se* and require more lengthy and specialized control measures.

Apart from the damage done to the population's health, influenza inflicts heavy economic losses. These losses result from people's temporary disability, post-influenza complications, mortality as a result of influenza as well as the population's additional morbidity and mortality. However, presently available estimates are only of economic losses resulting from the population's morbidity from influenza. In economic terms this includes payments for sick leave allowances, losses of working time and the cost of medical treatment for influenza-stricken patients. The major factor of economic loss is considered to be working time lost (up to 70%), with payments on sick leave allowances coming second (30% of the total losses), and medical treatment costs trailing behind at 2–3%. In 1976, the estimated economic loss attributed to the illness of one industrial worker amounted, on the average, to 128 roubles. The countrywide economic damage, done by influenza, reaches staggering figures. The economic losses, resulting from the incidence of the disease among gainfully-employed urban dwellers amounted to 1000–6000 million roubles, with 100–600 million roubles being added as a result of incidence among the non-working population (children, students and old-age pensioners); in some years the total losses exceeded 4000–6000 million roubles.

Thus, even a slight decrease in the population's morbidity may save the country millions of roubles. A radical decrease in influenza morbidity is, therefore, considered to be a major task of nationwide importance.

3. TRANSMISSION MECHANISMS OF INFLUENZA

An influenza patient is, in fact, the only source of influenza infection for susceptible people. Although a vast reservoir of influenza A viruses exists in nature among mammals and birds, these viruses are not pathogenic to humans, among whom they may only cause sporadic outbreaks, rather than epidemics. Emerging into the nasal secretions from damaged cells of the respiratory tract's villiferous epithelium, the influenza virus is released into the air in the form of multiple minute droplets during either physiological (breathing, speaking, crying or weeping) or pathological (sneezing, coughing) acts. No-one knows, to date, exactly how many infectious virus particles a patient releases, although the number obviously depends on the severity of the patient's catarrhal syndrome, as determined by the presence or absence of rhinitis, coughing and sneezing. Therefore, random infections are bound to play a less significant role in transmitting influenza.

The viral aerosol released by a patient, contains highly dispersed (5–25 μm) and low-dispersed (25 V–100 μm) particles, as well as small (100–250 μm) and large (over 250 μm) droplets. The amount of virus in an aerosol particle is determined by its volume — which is proportional to the cube radius of the particle. Therefore, the bulk of the influenza viruses which a viral aerosol contains is contained in large droplets. In assessing the role which different particles of a viral aerosol play in infecting humans, it is necessary to take their kinetics into account. As is known, the dispersal radius of a viral aerosol during coughing is about 1m. Droplets with a diameter of 1000 μm are dispersed to a distance of 11 m away from the source, those with a diameter of 100 μm to 1.1 m, and with a diameter of 10 μm to a distance of 0.13 m.

Having entered a recipient's respiratory tract, part of the aerosol is retained there, while the other part is removed with exhaled air. Particles over 10 μm in diameter are almost entirely retained in the upper respiratory tract. Half of the particles with a diameter of 3 μm settle in the upper respiratory tract and the other half in the lungs. In the case of 1 μm particles, about 88.5% are retained in the lungs and the rest in the upper respiratory tract. About 70% of 0.1–0.3 μm particles are removed with exhaled air.

'The structure of a viral aerosol undergoes rapid change primarily as a result of sedimentation, the rate of which is proportional to the square radius of a particle. Highly-dispersed particles settle at a rate of 0.00075–0.75 cm s, while low-dispersed particles settle at a rate of 1.9–27.0 cm s. On average, the dispersed phase of an aerosol settles at a rate of 30.5 cm min, the droplet phase at a rate of 30.5 cm s. The critical size at which droplets are no longer retained in a suspended state but settle is 0.1–0.2 cm, and the period of their float in the air is less than 3s. The aerosol structure is further subject to change because of rapid drying, which transforms it into a dispersed phase. Thus, particles with a diameter up to 25 μm dry up almost instantly (in less than 0.08 s) while it takes 0.31 s for particles with a diameter of 50 μm to dry up, 1.3 s for 100 μm particles, and 5.2 s for 200 μm particles. Thus, the fate of the droplet phase of an aerosol is first to settle down and then, after drying, to mix with dust. The highly-dispersed part of an aerosol is retained in the air for a long period of time and the particles are carried with the air over great distances. Meanwhile, aerosol concentration quickly diminishes as a result of sedimentation and mixing with dust. The influenza virus survives in the air for as long as 4 hours, serotype A strains being more resistant than those of serotype B. The virus survives in settled and dried aerosol droplets for 2 weeks on bed linen and as long as 5 weeks in domestic dust.

Since droplets on the major carriers of influenza infection and the period of activity of the aerosol's droplet phase is not long, and since infected particles are present right in front of a sick person within a radius of less than 1 m, it takes close (or direct) contact with such a person to catch the infection. The fact that the most frequent site of a pathological process among persons with influenza has been associated with the upper respiratory tract where the droplet phase of a viral aerosol is retained, is further proof of this conclusion. Since the likelihood of infection by inhaled aerosol is proportional to its concentration in the air and the duration of inspiration, while the concentration of highly dispersed aerosol is low and gradually decreases, the infection via highly-dispersed aerosol occurs seldom.

The results of random epidemiological surveys also point to the leading role of contact between humans in transmitting influenza infection. Thus, in the case for recipients with high susceptibility (most frequently the case with children) contact with a sick person with influenza would inevitably result in illness. Thus, it was shown that during influenza epidemics up to 90% of children in children's homes and orphanages fell ill after contacts with sick persons. Some 25% of children fell ill within 90 min of contact with a sick child, while 34.4 and 35.5% of children fell ill after 4–5h and 5.5–7h, respectively. In another survey it was found that in the course of 7 days of contact, the incidence of disease was lower than for contact of 8 days or more (22.2% against 30.0% in cases of influenza A; 19.4% against 31.2% in influenza B). Apart from the duration of contact, intensity also counts. If the beds of healthy children were placed closer than 3m from the beds of children with influenza, 42.1% (influenza A) and 38.6% (influenza B) of the children caught the infection, as against 30.4 and 29.0% morbidity when the beds were spaced over more than 3m apart.

The dependence of the probability of infection on the number of contacts with infection sources was also demonstrated: after contact with two or more children with influenza A and B, some 21.5 and 21.8% of the exposed children fell ill, while after contact with three sick children or more, 36.1 and 40.0% of the exposed children fell ill, respectively. Morbidity among secondary-school children during the three epidemics of influenza A/Hong Kong in 1969, 1970 and 1971 amounted to 29.9, 38.3 and 14.6%, i.e. was much lower compared to boarding school children (44.1, 43.3 and 33.6%) where the children are exposed to each other round-the-clock rather than only during classes.

As far as adults are concerned, the likelihood of catching the infection at one's place of work or within the family depends on the crowdedness of work places and on the living conditions in the household, including frequency and intensity of contacts. Thus, the higher the population density of work places, the higher the morbidity. Disease incidence is directly dependent on the size of the family, the number of public transport modes used, commuting time, frequency of visits to places of entertainment, etc.

A specific seasonal pattern of influenza and acute respiratory morbidity is also dependent on the specificity of contacts between people during different seasons. Sharp variations in morbidity as a result of influenza and acute respiratory illnesses (decrease in spring, rise in autumn) coincide with abrupt changes in the nature of contacts among people during these seasons. In spring and in summer people tend to disperse because of, say, annual leave, the end of the school year, and summer closure of pre-school children's institutions, which inevitably results in fewer contacts among people in closed premises — always potential breeding grounds for the spread of respiratory infections. In autumn, by contrast, most people are back from vacation, schools begin classes and pre-school children's institutions resume functioning. As a result, the number of contacts among people rises sharply.

It was found that morbidity figures resulting from influenza epidemics differ for different Soviet cities. This has primarily been the result of differences in their size of population, population density and the development of intra-city public transport. Thus, for instance, the highest morbidity (11.3%) due to influenza epidemics was found in cities with a population of 1 million and over. At the same time, in cities with a population from 500, 000 to 1 million, morbidity from influenza epidemics was 10.9%, and in towns with less than 500, 000 dwellers the morbidity level was 9.7%.

The use of mathematical models for predicting influenza epidemics has convincingly proved that contacts among people have been the key element in the spread of epidemics (Baroyan *et al.*). One of the major premises used for building a mathematical model of the spatial spread of influenza epidemics was the allegation that the infection spreads along public transport routes, and to substantiate this claim thousands of inter-city passenger traffic coefficients had to be analysed. Had this premise been incorrect, it would be virtually impossible to arrive at a good coincidence between the forecast and the actual morbidity.

4. PROPAGATION OF INFLUENZA EPIDEMICS

The history of influenza A epidemics shows that, as an epidemic disease, influenza has gained the widest circulation in the last 100 years (Ivannikov and Ismagulov, 1983; Zhdanov *et al.*, 1958; Smorodintsev and Korovin, 1961; Smorodintsev, 1984). As a result of growing transport communication throughout the world, influenza A caused epidemics which have recently become increasingly frequent, affecting all the world's population. In other words, it is possible to refer to it as a permanent influenza A pandemic, which can only be divided into individual episodes differing with respect to the antigenic characteristics of the epidemically important strains of the virus. Epidemics in individual countries and, moreover, in individual cities, constitute fragments of pandemics which originate in South-East Asia and the Pacific, from where they spread throughout the world by way of standard routes reflect existing international links. Until recently, with respect to the entire world there has never been a time which could be called a period between influenza A epidemics. Such periods could only be identified in relation to some specific areas; mankind as a whole represents a giant reservoir of the influenza A virus. It goes without saying that, in the case of influenza A, the world epidemic process develops as a result of the relevant influenza virus strains being brought by patients from epidemic countries into those countries which had so far been spared such epidemics.

With regard to influenza A, increasing intensification of the epidemic process has been observed throughout this century. Thus, it took approximately 18, 16 and 13 years for the strain with hemagglutinin H3 (which was in circulation about the year 1900) to be replaced by the influenza virus with hemagglutinin Hsw in 1918, and then to be further replaced first by the virus with hemagglutinin H0 and then with hemagglutinin H1. Later on, the next epidemically important variants of the virus (which differed radically from each other by their hemagglutinin) appeared every 10–11 years; the Hong Kong influenza virus has proved to be a veritable "chameleon" for, since 1971, it has presented a new variant every year. Since the selective effect of community immunity to influenza does not require any further proof, the increasing rate of antigenic changes is evidence of growing community immunity to influenza A which, in turn, constitutes a function of mankind's growing pro-influenza disposition — a fact testifying to the intensification of the epidemic process. The most intensive pandemics have been associated with the greatest upheavals in mankind's history. Thus, the influenza A(HowN1) pandemic of 1918–1920 occurred after World War I, the influenza A(H1N1) pandemic of 1946–1948 occurred after World War II and the pandemics of influenza A(H2N2) of 1957 and influenza A(H3N2) of 1968 developed at a time of a revolution in science and technology and of rapidly growing interstate relations. The rates of pandemics development have also changed greatly: if the pandemic of swine influenza lasted 3 years, and in the 1940s the pandemic of influenza A1 continued for 2 years, then a rather weak pandemic of influenza A1 in 1977–1978, which mainly affected children and young people, swept the entire world in a single year. The frequency of influenza epidemics has also increased. Thus, if before the arrival of Asian influenza in 1956, influenza epidemics in Leningrad occurred, on average, every third year, since 1957 their frequency has increased to 18 months, and since 1977 epidemics take place every year and sometimes even twice a year (Table 1).

The abrupt intensification of the epidemic process and the associated increase in naturally acquired specific resistance to influenza have probably reached their peak. This situation has significantly changed the nature of influenza propagation throughout the world in recent years. Cases in point are the epidemics of influenza A(H3N2) and A(H1N1). If before 1977 every influenza epidemic in the Soviet Union constituted a fragment of the pandemic, nowadays there occur intervals, lasting as long as several months, with neither any mention of influenza epidemics anywhere in the world, nor even any information about virus isolates.

In contrast to influenza A, it is impossible to mention a single influenza B epidemic which developed into a pandemic over a similarly brief period of time. In the course of one year influenza B propagates most intensively in some countries, or sometimes through a continent — which more often than not is Europe. Such epidemics, as a rule, are polyetiological, that is, two or more variations of influenza B virus could be simultaneously identified. The old strains are replaced by new ones very slowly, and antigenic changes in influenza B viruses occur quite seldom. Thus, for instance, only two changes in the epidemically relevant viruses of influenza B took place during the 1967–1977 period. There may be quite prolonged intervals (as long as a year) when pronounced epidemics have not been registered in a single country.

If numerous influenza A virus isolates are proof of a developing epidemic, and there are practically no influenza A virus isolates in the inter-epidemic period, influenza B virus may be isolated in many cases during occasional outbreaks which, nevertheless, very seldom develop into epidemics. In other words, unlike influenza A, during an inter-epidemic period influenza B is characterized by the continuous circulation of viruses among the population (almost exclusively among children).

Influenza B epidemics occur much more seldom than those caused by influenza A. Thus, from 1935 to 1979, 14 influenza B epidemics were recorded in Leningrad, i.e. one epidemic in approximately four years. In some cases such epidemics affected only children. In recent years, however, influenza B epidemics have become more frequent, with higher morbidity. This again, as in the case with influenza A, indicates an intensification of the epidemic process as a result of more frequent contacts among people.

The peculiarities of the worldwide propagation patterns of influenza B suggest the relative independence of their origins and propagation in different countries. From this point of view, influenza B epidemics may be described as predominantly endogenic, arising in the course of a local epidemic process, but having a low capacity for rapid generalization over large areas.

5. COMMUNITY IMMUNITY AND THE EPIDEMIC PROCESS

The retention for a long period of anti-hemagglutinins which are specific for certain strains of influenza viruses have been long known from the work of Francis, who was the first to discover antibodies to swine influenza in elderly people. The presence of antibodies to this virus in the blood of elderly people was subsequently repeatedly confirmed. That the retention of humoral immunity to influenza for scores of years is not incidental has been confirmed by finding antibodies in elderly people, first to hemagglutinin H2 and then to hemagglutinin H3. In the blood sera collected in October 1977, antibodies to hemagglutinin Hsw were found in persons older than 50 years, to hemagglutinin H0 in people older than 40 years, and to hemagglutinin H1 in people older than 30, that is, in people who could have been infected 20 to 50 years previously with the corresponding influenza virus. Anti-hemagglutinins H2 and H3 were also found in all persons examined. It is interesting to note that the geometric mean of antibody titres to influenza A/Singapore/1/57 (H2N2) virus, which circulated 20 years ago, was 1:30 and higher in all groups.

It has been recently proved that antibodies to neuraminidase are retained in the blood for the same long period of time.

Until recently, the retention of strain-specific antibodies was regarded as a trace reaction, similar to partial immunological memory, which fails by itself to provide effective protection. However, the events associated with the 1977 epidemic caused by influenza A(H1N1) virus — a virus similar to the influenza A1 which had circulated 20 years before — refuted this interpretation. It turned out that the 1977 epidemic affected mainly people who had never been infected with this virus since all of them were born after 1957, the year when influenza A (H1N1) virus ceased to circulate. Thus, morbidity among schoolchildren aged 7 to 14 years and among teenagers amounted to 21.6 and 20.1%, respectively: a 7 to 9-fold increase over the morbidity among people aged from 35 to 60. Thus, strain-specific immunity to influenza may survive either for the course of one's life or for at least dozens of years; it may protect against disease resulting from infection with a closely related influenza virus.

What still needs to be clarified is how effective the crossed immunological protection is in the case of influenza A. One should recall, first of all, that only a small fraction of the population contracts influenza during an epidemic. In the majority of cases, influenza morbidity in adults is less than 10% and never above 23%. Some researchers have been looking for an explanation of this phenomenon in the possibly innate or non-specific immunity of most people to influenza virus.

However, observations over a cohort conducted during four influenza A epidemics in the 1968–1971 period show that the proportion of people who proved to be immune to influenza A(H3N2) amounted to a mere 2–3%. Here it should be recalled that when some islanders who had been isolated from the rest of the world for some time encountered influenza viruses for the first time ever, their susceptibility was found to be very high. This fact once again testifies that man's susceptibility to influenza is generally high.

One may presume the existence of crossed immunological effects in the course of influenza. First, there are the results of tests involving volunteers immunized with live influenza vaccine. These tests proved that protective vaccines are possible against post-immunization infection not only with homologous, but also with heterologous influenza A virus. Similar data were obtained in the course of observations of persons who were immunized with inactivated influenza vaccine but in the course of a natural epidemic.

Moreover, analysis of data obtained after years of studying the epidemiological efficacy of live influenza vaccines in the Soviet Union, showed that anti-influenza immunization considerably decreases the population's morbidity although, as is well known, vaccine strains have never completely corresponded in terms of their antigenic structure with circulating viruses. Tests of the epidemiological efficacy of live influenza vaccines in the course of abrupt changes in the antigenic structure of viruses deserve special mention. It turns out that even in these conditions morbidity among those vaccinated was, on average, some 33% lower than among the controls.

Finally, morbidity in the course of any influenza epidemic depends on the morbidity of preceding epidemics. And although the morbidities of the most closely spaced epidemics correlate poorly with each other, epidemics that took place long ago, do have a considerable impact on the current epidemic, which confirms the conclusion about the continuing effect that heterologous anti-influenza immunity has on morbidity.

Therefore, one may consider it as established that strain-specific immunity to influenza A may be retained for dozens of years and will protect against disease caused by homologous virus infection. The existence of a sufficiently broad spectrum of specificity in post-influenza immunity has been proved beyond any doubt. This probably ensures protection against the disease where immunity has been discovered to closely related variants of influenza A virus. Also important in the case of influenza A, are the heterologous effects of regulating the epidemic process. The conclusions reached make it possible to view with great optimism the prospects for influenza prophylaxis by the use of vaccines. Meanwhile, the mechanisms of retention of post-influenza immunity over a long time, and the degree of its heterogeneity, should be given top priority in research programmes.

6. MODERN THEORY OF THE INFLUENZA EPIDEMIC PROCESS

Summing up the conclusions arrived at above, and interpreting the epidemic process in the light of the latest data available, allows one to present the basic premises of a modern theory of the influenza epidemic process in a nutshell. This theory unites within a single system all aspects of influenza epidemiology and explains the trends and mechanisms by which the influenza epidemic process alters as a result of changes in the relative importance of the virus, the mechanism of infection transmission and community immunity. In the past any change in the influenza epidemic process was regarded as a result of antigenic changes in the influenza virus.

Although the roles of the mechanism of infection transmission and community immunity have been already discussed, the major results should be repeated here. The leading role in changing the nature of the epidemic process during influenza belongs to the varying intensity of contacts among people; it can be said with certainty that influenza infection is mainly transmitted in the course of direct contacts between patients and susceptible people. The history of the propagation of influenza epidemics shows that changes in the nature of these epidemics have been caused by the growing urbanization of man's life — a process which entails an increasing number of contacts among people. The role of this social factor (contacts between people) has been confirmed by the decisive impact of such indicators as the size and density of the population, its morbidity in different areas and urban centres, as well as by the seasonal patterns of influenza and acute respiratory diseases. It has been established that, in terms of the organism's general resistance, the diversity of natural and climatic factors plays a secondary and minor role. The key role of contacts between people in the propagation of influenza has been confirmed by the successful practice of forecasting the development of influenza epidemics over large areas using mathematical models. Such models are based on the premise of epidemic propagation along transport routes.

It is important to note that, like the human population, the population of influenza viruses is not fixed and homogeneous. It is common knowledge, since it is always mentioned in every book on influenza and widely researched, that influenza virus exhibits tremendous variability. The never-ceasing nature of the worldwide epidemic process affecting man, manifesting itself as influenza epidemics occurring in one area or another, requires consideration from a fresh perspective. The permanent nature of epidemics ensures the survival of a numerous virus population within which its antigenic structure keeps changing all the time. This refutes the hypotheses, current for many decades, concerning the simultaneous occurrence of epidemics over large areas, the disappearance of influenza viruses after epidemics and their persistence until a new epidemic. The simultaneous discussion of all existing hypotheses concerning the shift variants of influenza A viruses (the hypothesis of recombination of human and animal influenza viruses, the hypothesis of influenza virus persistence in the human organism, the hypothesis of the unceasing circulation of all virus variants among men, the mutation hypothesis) makes it possible to show that none contradict the existing epidemiological data, but none of them has any advantage over the others. On the other hand, it is clear that not a single existing hypothesis concerning the origins of epidemic influenza viruses allows one to elaborate a method of long-term epidemiological forecast. This emphasizes the importance of the premise on the possible limitation of epidemically-meaningful antigenic variants of influenza A viruses for long-term etiological forecasts using immunological methods.

The role of a dynamically changing and growing herd immunity to influenza among people has not yet been fully assessed. At present it may be stated with confidence that post-influenza immunity exists and is retained for dozens of years, and that there are wide-ranging immunological crossed interactions among different serotypes of influenza A viruses which have a role to play — all this contrary to the earlier view of the weakness of anti-influenza immunity. The premise about the impossibility of giving separate definitions to the concepts of influenza virus virulence and of immunity to influenza, has been substantiated; it has also been shown that, in the case of influenza, these two notions may only be defined jointly and simultaneously, one through the other.

The theory of the influenza epidemic process allows one to analyse and to forecast the outcome of the present epidemiological situation as far as influenza is concerned. An analysis of the worldwide spread of influenza epidemics has revealed a marked decrease in their intensity and propagation since about the end of 1976. If before that date each variant of influenza A virus, including drift variants, used to acquire pandemic proportions with several countries registering a marked epidemic virtually every month, in recent years the worldwide epidemic process has become intermittent: in some months, or even several months in a row, there is no information either about new influenza epidemics or even about influenza virus isolates.

A more detailed comparison of previous epidemics with the present epidemiological situation with respect to influenza A, and materials describing influenza A epidemics in the Soviet Union, show that such epidemics are assuming a markedly endogenic type of propagation. The typical routes of epidemics propagation across the country have been disrupted; the onset of epidemics is now reported in towns where they had never sprung up before, towns, moreover, which often have not developed international links. Furthermore, in the majority of cases influenza no longer spreads to all towns, as before, but rather to a few of those towns under surveillance. In some towns the dynamics of the epidemic are rather haphazard: the epidemics may be either very brief (2–3 weeks) or very prolonged. Epidemics may from the very start show maximum morbidity or, conversely, may terminate with high morbidity. Or there may be epidemics without marked morbidity peaks but exhibiting a constant moderate level of morbidity slightly exceeding the epidemic threshold in the course of several weeks. The total duration of epidemics in the country has decreased from 15 weeks to 12.6 weeks; this testifies to the more sluggish development of influenza epidemics in recent years. One index (showing the number of towns affected by epidemics in one week) is a good illustration of this trend: it has declined twofold from 6.6 to 3.1 towns. The morbidity indicators also point to a sharp decrease in the intensity of influenza A epidemics. Thus, the maximum morbidity per epidemic week decreased from 3.2 to 1.3% while prevalence over the entire epidemic dropped from 11.1 to 4.1%.

In recent years, influenza B epidemics have intensified. Thus, if in the past influenza B epidemics attacked, on average, 49% of the towns under surveillance, recent years saw this figure go up to 97%. The last three influenza B epidemics affected all the towns under surveillance, and their duration increased from 6.5 to 8.5 weeks. Morbidity during the epidemic peak week increased from 1 to 1.9%, and prevalence more than doubled from 4.2 to 9%. It should be emphasized that the incidence has especially risen among children. This applies even to influenza A epidemics, although prevalence during these epidemics, as already stated, declined. The situation is very clear in the analysis of the Leningrad population's morbidity broken down by age (Table 1). In recent years morbidity in children aged 0–2, 3–6 and 7–14 has reached record figures which can only be compared with 1970, the year when there was a mixed epidemic of influenza A and B.

These changes in the intensity of the epidemic process for influenza A and B take place against a background of a marked rise in the population's morbidity as a result of acute respiratory infections in recent years. Thus, whereas in 1958–1962 the average annual indicator of the population's acute respiratory morbidity was 12.6% throughout the country, it went up to 22.3% in 1978–1982. This rise in morbidity has been due mainly to the increasing incidence of acute respiratory infections of a non-influenzal etiology. These illnesses now constitute 75% and over of the morbidity resulting from influenza and acute respiratory illnesses. It is important to note that children contribute more than half of the total morbidity.

The declining intensity of influenza A epidemics, the growing intensity of influenza B epidemics and the increasing incidence of acute respiratory illnesses of a non-influenzal etiology are far from accidental and can be attributed to common causes. The prime reason for all these changes lies in the intensification of the transmission mechanism of influenza and other acute respiratory diseases as a result of the increasing frequency of contacts among people all over the world, especially in recent decades. The intensification of the virus transmission mechanism results in a rise in morbidity, more frequent and severe epidemics, more frequent antigenic changes in influenza A viruses, and the simultaneous circulation of several antigenic variants. One witnesses the growing "epidemicization" of the population and the related increase in community immunity, as well as a rise in the "quality" of immunity because of the simultaneous circulation of different virus serosubtypes and their variants for influenza A. When community immunity reaches a particular sufficiently high level, a decline in the intensity of the epidemic process takes place – influenza A being a case in point. As far as influenza B and other acute viral respiratory illnesses are concerned, immunity has not yet reached saturation point, so that morbidity due to these infections is on the rise. This increase, however, cannot last indefinitely, and there must come a moment of equilibrium between community immunity and the intensification of the virus transmission mechanism. Nor can the intensification be infinite; there are, for instance, trends pointing to saturation of the growth curve of transport routes. In the coming years, therefore, one may expect the intensity of influenza A and B epidemics to stabilize roughly at the present level. Very soon, morbidity due to other respiratory infections should also reach stable levels and stop increasing. This forecast assumes that influenza and other acute respiratory illnesses will still be uncontrollable infections. And the possibility should not be overlooked that a certain temporary increase in the intensity of influenza A epidemics is possible if an absolutely new pandemic variant appears.

REFERENCES

Baroyan, O.V., Rvachev, L.A. and Ivannikov, Yu.G. (1977) Simulating and Predicting Influenza Epidemics for the USSR. Moscow:The N.F. Gamaleya Institute of Microbiology and Epidemiology (In Russian).

Ivannikov, Yu.G. and Ismagulov, A.A. (1983) The Epidemiology of Influenza. Alma Ata:Kasahstan (In Russian).

Martynov, Yu.S. (1970) Damage to the Nervous System in Influenza and Influenza-like Illnesses. Moscow: Meditsina (In Russian).

Smorodintsev, A.A. and Korovin, A.A. (1961) Influenza. Leningrad: Medgiz (In Russian).

Smorodintsev, A.A. (1984) Influenza and its Prevention. Leningrad: Meditsina (In Russian).

Zhdanov, V.M., Soloviev, V.D. and Epstein, F.G. (1958) Conception of Influenza. Moscow: Medgiz (In Russian).

Zlydnikov, D.M. and Smorodintsev, A.A. (1974) Acute Respiratory Diseases. Leningrad: Meditsina (In Russian).

Sov. Med. Rev. E. Virology Reviews, Vol. 3. 1990, pp. 117–129
Photocopying permitted by license only

A MODERN SYSTEM FOR COMPREHENSIVE PREVENTION OF INFLUENZA IN THE USSR

A.S. SHADRIN and G.I. KARPUKHIN

All-Union Research Institute of Influenza, USSR Ministry of Health, Leningrad

Abstract

Since 1980, from 30 to 50% of the population in a number of Soviet cities have been immunized annually with inactivated or live vaccines. Antiviral chemoprophylactic drugs (rimantadine and Oxolyn) and immunological preparations based on human blood (leukocytic interferon and anti-influenza immunoglobulin), have been used as emergency preventive measures during influenza epidemics. In test observations the effectiveness coefficient ranged from 24.0 to 30.7% for inactivated vaccines, from 7.6 to 32.4% for live vaccines, and from 0 to 77% for rimantadine. A system of influenza control measures has resulted in an 11.4 to 19.2% decline in morbidity against the expected percentage, or in at least 50,000 fewer cases per 1 million population.

Key words: vaccines; chemoprophylaxis; efficiency

1. INTRODUCTION

Four major strategies for influenza prevention have been developed in the world today. (Shadrin *et al.*, 1982). In the USA and the majority of European countries influenza prophylaxis is mainly oriented towards a 'high risk' group in whom unfavourable complications of the disease are likely (elderly people, chronically ill adults and children). The total numbers vaccinated annually amount to 5–10% of a country's population.

In Japan, the priority contingent of the population to whom influenza-preventive measures are applied, are children (Dowdle *et al.*, 1980). Each year 80% of children and teenagers aged from 3 to 18, who attend either children's institutions or schools, are immunized as well as 3% of the adults belonging to the 'high risk' group. All in all, some 20% of the country's population are covered by the annual immunization programme.

For a long time (since 1954) the anti-influenza campaign in the Soviet Union has mainly consisted of selective immunization of healthy adults with a live vaccine. However, less than 10–15% of the population were immunized annually in the majority of towns and urban settlements. What is striking, when assessing the efficiency of influenza prophylaxis, is a glaring discrepancy between the positive results of using individual drugs in the course of limited epidemiological observations, on the one hand, and the lack of appreciable results from the same drugs in large-scale practical applications. The morbidity of influenza and other acute respiratory diseases (ARD) has been steadily rising (Karpukhin, 1986). Whereas in the 1960s in the Soviet Union about 1500 cases of influenza per 10 000 were recorded annually (from July to June), in the 1970s this number was slightly above 2000, and over 2500 in the 1980s. Moreover, there are no grounds for believing that this rise in morbidity was caused by ARDs other than influenza, since, according to laboratory data, the latter's share within all ARDs has not declined (Karpova *et al.*, 1983; Gagarinova, 1983).

2. A SYSTEM OF COMPREHENSIVE PREVENTIVE MEASURES

A new system, which would be the fourth on our list, has been developed at the All-Union Influenza Research Institute in Leningrad (Karpukhin, 1981, 1983, 1986; Karpukhin and Shadrin, 1977). It presupposes the most rational application of anti-influenza drugs. The system was tested in the town of Severodvinsk (Gagarinova *et al.*, 1982, 1983a and b; Ivanova and Gagarinova, 1981) and since 1980 it has been introduced in several other cities (Shadrin and Karpukhin, 1981; Shadrin and Zykov, 1984; Shadrin *et al.*, 1982, 1983, 1985). The system is comprehensive in terms of ways and means of application over large areas and in regard to the groups of the population being protected.

The idea of an epidemic process as a self-regulating ecological system in which populations of the host and the parasite (Belyakov *et al.*, 1979; Shadrin and Golubev, 1980) — heterogeneous towards each other — interact, represents the theoretical basis of the new system of measures. All those residing in a certain territory are regarded as a host population within which the parasite, or the influenza virus, circulates (Karpukhin, 1986; Shadrin and Zykov, 1984).

In the host population with its high herd immunity, i.e. with a significant number of completely or partially non-responding individuals, changes in pathogen ecology occur in terms of the prevalence of low-virulence strains capable of causing only mild sporadic diseases. The macroorganism would seem more tolerant to such strains than to virulent ones, and hence it does not alert all available defense mechanisms. This increases the probability of the virus surviving as a biological species under unfavourable habitat conditions. The pathogen's low-virulence strains have but a slim chance of crossing the immunity barriers of vaccinees even when these barriers are relatively weak.

The situation is quite different, however, when the bulk of the host population is not immune but is susceptible to the virus. In such cases highly virulent strains, capable of overcoming not only the relatively weak post-vaccination defences, but natural post-infection immunity as well, may form.

The ecological patterns of influenza epidemics are clearly manifested in cases where the host's immunity has been ensured by means of non-specific methods of protection, e.g. antiviral drugs. Thus, according to Karpukhin *et al.* (1979), when only 30% of all staff workers took rimantadine as a preventive during an influenza epidemic, the likelihood of contracting the disease declined by some 33% compared to controls, by 66% when half of the staff took rimantadene, or even by 90% with three-quarters of the staff taking the drug.

The concept of an epidemic as a system, which involves populations rather than isolated individuals or persons, correctly and clearly orients anti-epidemic measures towards the realization that influenza may be effectively controlled at the populational level only. In order effectively to protect the population against influenza by using the drugs presently available (drugs which do not provide absolute immunity) it is necessary to protect the host population as a whole, i.e. a population as a territorial entity, for instance, a large city or a region, rather than individuals or separate groups. Under some specific circumstances effective influenza control is possible within a restricted area, for instance, in relatively isolated army units (Meiklejohn, 1983, Romanov *et al.*, 1982), which in this particular case may be regarded as host micropopulations.

The populational–ecological approach to influenza control cannot at present be put into effect using only one or a group of drugs, nor by means of vaccines alone. Therefore, apart from the large-scale application of anti-influenza measures, other major principles of the system, such as the comprehensive nature of these measures, as well as the use of all effective means must be considered. This will mean a differential application of anti-influenza drugs for different population groups or under different conditions.

3. USE OF VACCINES FOR INFLUENZA PROPHYLAXIS

The new tactic of protecting the Soviet Union's population against influenza rests on the use of inactivated or live influenza vaccine (IIV and LIV). The anti-viral drugs (rimantadine and Oxolyn) as well as immunopreparations based on human blood (leukocytic interferon and anti-influenza immunoglobulin) (Table 1) have to play an ancillary role. A subunit IIV (Bashlyaeva *et al.*, 1986), LIV for children (Polezhayev *et al.*, 1983; Alexandrova *et al.*, 1984), and other drugs are presently at different stages of trial and development.

Soviet-made IIVs differ from foreign-made vaccines by a lower concentration of virus protein per vaccine dose (3–4 μg of hemagglutinin of every virus) since they are intended for mass-scale immunization by intracutaneous injection by means of the most convenient jet injectors. It is estimated (Bykov, 1980) that in order to prevent an influenza epidemic in a town it is necessary to inoculate about 80% of its inhabitants, provided the vaccine reduces the probability of contracting influenza by a factor of two. The experience of the town of Severodvinsk shows that even with the vaccination team doing its utmost only 70% of the population could be immunized (Gagarinova *et al.*, 1983a), while vaccine efficacy was on average below the required level mentioned above. Therefore, although it has not so far been possible to completely arrest the development of influenza epidemics anywhere, it is still quite realistic to expect to curtail them significantly (Gagarinova *et al.*, 1982, 1983a and b).

The safety and efficacy of anti-influenza vaccines intended for mass-scale immunization programmes, were checked annually by the double blind method. The reactivity of both IIV and LIV was well below the prescribed 2–3% level (by general reaction, with body temperature rising above 37.5°C). According to several years' observation of the somatic disease incidence in persons who had been annually immunized, no damaging effects of vaccination on the human organism have been found (Maksakova *et al.*, 1985).

The antigenic activity of the whole-virion IIV to viruses of the A(H3N2) type was steadily high and included a four-fold increase or

Table 1 Major anti-influenza agents for mass-scale application.

Agent	*Activity*	*Administration*	*Protected groups*
Vaccines:			
Inactivated whole virion	3–4 μg HA of each serotype virus in injection dose	Injection i.c.; one per year	Adults and children (over 11 years)
Inactivated whole virion + additional purification	As above	Injection i.c.; one per year	Children (7–10 years)
Inactivated split-product absorbed	As above	Injection s.c.; one per year	Adults and children (over 11 years)
Live vaccine	10^5–10^6 EID of each serotype virus in injection dose	i.n.; two vaccinations per year	16–55-year-old old adults (16–55 years)
Chemical agents:			
Rimantadine	0.05 g	p.o., daily during epidemic	Adults (over 15 years)
Oxolyn	0.25% ointment	i.n., twice daily during epidemic	Adults and children
Immunological agents:			
Leukocytic interferon	32–64 unit	i.n.; twice daily in epidemic foci	Children (below 7 years)
Anit-influenza immunoglobulin (donor) human	1:160–1:320 titre in HI test	Injection i.m.,; 1/mL, in epidemic foci	Children (below 3 years)

HA = hemagglutinin; HI = hemagglutinin - inhibition i.c. intracutaneous; s.c. = subcutaneous; i.m. = intramuscular; i.n. = intranasal; p.o. = orally.

more in the anti-hemagglutinin titre in 53.5–87·0% of primary vaccinees with a low pre-vaccination antibody titre (⩽1:20). Considerable changes in the seroconversion rate from 31.9–40.0% in the fall of 1983 and 1984, up to 64.7–77.0% in other years were recorded with A(H1N1) viruses (Table 2). The reasons for the lower immunogenicity of the A(H1N1) component in 1983–1984 are not clear (Shadrin *et al.*, 1985; Maksakova *et al.*, 1985; Shadrin and Karpukhin, 1981; Ignatieva *et al.*,

Table 2 Immunogenicity of influenza vaccines.

Year	Test virus	IIV (divaccine)[a] sero-conversions (%)[c]	IIV (divaccine)[a] persons with 1:40 titre[d]	LIV (monoagents)[b] sero-conversions (%)[c]	LIV (monoagents)[b] persons with 1:40 titre[d]
1980–1981	A(H1N1)	77.0	79.4	12.0	16.1
	A(H3N2)	65.0	61.8	28.9	38.7
1981–1982	A(H1N1)	73.3	67.4	17.9	21.6
	A(H3N2)	87.0	85.7	29.3	24.7
1982–1983	A(H1N1)	64.7	57.3	14.3	15.9
	A(H3N2)	53.5	45.6	14.3	16.0
1983–1984	A(H1N1)	31.9	38.9	—	—
	A(H3N2)	66.7	82.7	21.9	33.8
1984–1985	A(H1N1)	40.0	62.5	—	—
	A(H3N2)	81.0	81.3	30.8	78.8

[a] A(H1N1) + A(H3N2); [b] A(H1N1) or A(H3N2); [c] persons with pre-vaccination antihemagglutinin titre 1:20; [d]antihemagglutinin titre one month after vaccination.

1983). LIV immunogenicity was not high and seroconversions were observed in 12.0–30.8% of vaccinees with a low pre-vaccination antibody titre. Nasopharyngeal secretions of people vaccinated with LIV, revealed a reliable increase in the specific antibody titre in 90–95% of cases. Inactivated influenza vaccines did not stimulate the production of secretory antibodies. In order to assess the epidemiological (prophylactic) efficiency of anti-influenza immunization, it is necessary to specify that actual virus strains of the A(H1N1) and A(H3N2) serosubtypes were used as a divaccine both in the observed tests (Table 3) as well as during mass-scale vaccinations. There is, however,

Table 3 Epidemiological efficiency of influenza vaccine in observed tests.

Year	Epidemic type	Morbidity (%) IIV[b]	Morbidity (%) LIV[c]	Morbidity (%) Placebo	EC (%)[a] IIV[b]	EC (%)[a] LIV[c]
1980–1981	B + A(H1N1)	25.5	32.7	38.3	33.4	14.6
1981–1982	A(H1N1)	7.0	9.5	10.1	30.7	5.9
					(41.2)	(37.5)
1982–1983	A(H3N2)	11.2	9.2	14.7	23.8	37.4
1983–1984	B + A(H1N1)	17.6	23.1	25.0	29.3	7.6
1984–1985	A(H3N2)	12.7	12.9	16.6	24.0	27.5

[a] Effectiveness coefficient with ratio of influenza A after the serological correction of diagnosis in parentheses; [b] divaccine A(H1N1) + A(H3N2); [c] monovaccine A(H1N1) in 1983–1985 (two seasons); and divaccine A(HINI) + A(H2N2) in 1980–1983 (three seasons).

inconsistency with respect to LIV: if monovaccines of the A(H1N1) serosubtype were used in the fall of 1980–1982, then in 1983–1984 it was the A(H3N2) serosubtype that was utilized. In 1980–1983, in the course of observed tests live divaccines A(H1N1) + A(H3N2) were used, while monovaccine A(H3N2) was applied in 1983–1985. Thus, during the observed tests, the strain composition of IIV and LIV corresponded to the epidemiological situation during three seasons out of five (except 1980–1981 and 1983–1984).

Given the total number of cases of influenza and acute respiratory illness, the effectiveness coefficient (EC) of IIV ranged from 23.8 to 33.4%. Taking into account the serological diagnosis, the likelihood of influenza A developing among vaccinated subjects decreased by 41.2% during the observed tests of 1981–1982.

The efficacy of LIV was comparable with that of IIV during the influenza epidemics caused by viruses of the A(H1N1) and A(H3N2) serosubtypes. During epidemics of mixed etiology with the predominance of influenza B virus, the preventive effect of LIV proved insignificant (EC = 7.6–14.6) (Shadrin *et al.*, 1985; Shadrin and Karpukhin, 1981; Ignatieva *et al.*, 1983, 1986).

4. EMERGENCY INFLUENZA PROPHYLAXIS

Whenever it has been used for mass-scale application, rimantadine proved the most efficient and useful agent for emergency prophylaxis (preventive treatment) of influenza (Shadrin *et al.*, 1982, 1984; Yerofeyeva *et al.*, 1982).

The scale of application and the efficacy of other chemoprophylactic and immunological agents (see Table 1) were much inferior to rimantadine (Karpukhin, 1986), and, therefore, they are not reviewed here. Karpukhin and Shadrin (1977) proposed to subdivide emergency prophylactic measures into two kinds: focal prophylaxis and planned emergency procedure, which differ fundamentally in their organization and duration, and in the population groups involved. Focal prophylaxis is conducted in epidemic foci among people who have had direct contact with an ill person in the family, a hostel (barracks), or a hospital ward. The course of treatment varies from 2 days, if contact with the source of infection had been immediately discontinued, to 5–7 days if the infected person had not been isolated and contact with him had not been cut off. Planned emergency protective procedure covers non-vaccinated subjects belonging to a defined high-risk group susceptible to illness during influenza epidemics. Planned prophylactic treatment with a daily 0.05 g rimantadine decreased by 53.0–77.4% the likelihood of illness

Table 4 Epidemiological efficiency of rimantadine.

Epidemic type	*Year*	*Number of test subjects*	*Morbidity (%)*		*Efficiency (%) coefficient*
			Test subjects	*Control*	
A(H1N1)	1982	986	1.3	3.6	63.9
A(H3N2)	1980	1352	3.7	8.5	56.5
A(H3N2)	1983	507	3.1	6.6	53.0
A(H3N2)	1983	654	2.6	11.5	77.4
A(H3N2)	1985	981	19.4	13.5	−43.7
B	1981	348	13.4	13.5	0.7
B	1984	1296	11.4	14.2	19.7

during influenza epidemics caused by viruses A(H1N1) and A(H3N2) (Table 4). The only exception was a test conducted in 1985, when rimantadine was found to be ineffective, probably due to the fact that highly rimantadine-resistant strains of influenza virus serotype A(H3N2) had been isolated both prior to and during drug taking by a group of subjects staying in a sort of boarding house. During the epidemics, caused mainly by influenza B virus, the effectiveness coefficients were low (0.7–19.7).

Focal emergency prophylaxis of infuenza reduced the number of secondary cases of influenza A in families by a factor of 6 to 10 (Karpukhin and Shadrin, 1977; Karpukhin, 1986).

5. ASSESSMENT OF THE EFFICIENCY OF COMPREHENSIVE PREVENTIVE MEASURES

The complex of influenza measures designed to prevent influenza consists of four elements:

1. vaccination of the population in the pre-epidemic period;
2. emergency prophylaxis of illness by means of chemical and immunological anti-viral agents during epidemics;
3. early etiotropic treatment of sick persons;
4. general anti-epidemic measures.

In assessing the efficiency of measures in towns, the most demonstrative results were obtained in Severodvinsk in 1979–1980 (Gagarinova *et al.*, 1982, 1983a and b). The epidemic-related increase in influenza and other acute respiratory morbidity (epidemic increment) proved to be 2.7 times less than expected, in comparison with another 44 towns where the new influenza control system had not yet been introduced. The average duration of incapacity in uncomplicated influenza cases declined by 1.2 days, and in case of complications by 3.3 days. The proportion of influenza A in the pattern of acute respiratory diseases dropped from 37.1 to 18.7. The number of cases of pneumonia and bronchitis, the most frequent post-influenza complications, declined 1.8 and 1.4 times, respectively.

Table 5 summarizes the result of influenza-control measures during the 1980–1985 period. These are also reviewed elsewhere by Shadrin and Zykov (1984) and Shadrin *et al.* (1985). The effectiveness coefficients of these measures (EC) were calculated by the formula:

$$EC = (1 - (M_{c'} : M_{t'}) : (M_c : M_t) \times 100$$

where: M_t is the morbidity index in a town in 1980–1985 ("test"), M_c is the similar index for the country ('control), $M_{t,}$ and $M_{c,}$ are morbidity indices for 1968–1980, before the implementation of the measures.

In Table 5, towns are listed by their average annual percentage of vaccinees. Other anti-influenza agents were used, as a rule, more widely in towns with a high percentage of vaccinees (Shadrin *et al.*, 1985; Shadrin and Zykov, 1984).

There is a clear correlation between the effectiveness coefficient of the measures taken and the percentage of vaccinated subjects ($r = 0.63$).

On average, in towns where more than 50% of the population were vaccinated, the coefficient was 19.2, while in other groups it ranged from 5.2 to 11.4. Since influenza A accounts for less than 30% of acute respiratory diseases, a decline in the average annual indicators of total influenza and ARD morbidity by 10–20% against the expected level is probably close to the calculated limit, i.e. to the full realization of the potential efficiency of available mass methods of influenza prophylaxis.

Given the present-day level of acute respiratory morbidity (40–45 cases per 100 town dwellers), its decline by 11.4–19.2% (Table 5) means that in some towns the average number of patients per one million inhabitants decreased, on average, by at least 50 000 cases, while influenza-related economic losses decreased by 2.5–3.0 million roubles.

The morbidity indicators, without a breakdown by age groups, do not provide a full picture of the efficiency of influenza protective measures, which were extended only to adults over the observed period (1980–1985). In the majority of towns there was, unfortunately, no separate registration of sick adults and children. The data available for some cities (Moscow, Leningrad and Gorky) show that over this period adult morbidity declined by 10–20%, while that of children increased steeply.

Table 5 Efficiency assessment of anti-influenza measures in some cities for the 1980–1985 period.

	Average annual indicators			
City	*Population vaccinated (%) 1980–1985*	*Influenza and ARD morbidity (per 100) 1968–1980*	*Influenza and ARD morbidity (per 100) 1980–1985*	*Effectiveness coefficient[a] (%)*
Rostov	58.3	35.6	34.4	17.3
Minsk	55.6	54.7	56.4	13.8
Kharkov	53.7	59.0	59.9	14.5
Severodvinsk	53.6	40.3	31.3	34.6
Perm	50.8	37.4	35.6	19.3
Frunze	49.8	28.4	28.2	16.0
Average indicator	53.6	42.6	40.9	19.2
Lvov	48.7	41.0	47.3	2.0
Sverdlovsk	47.2	35.6	32.4	23.7
Irkutsk	44.8	28.8	28.9	15.2
Odessa	43.6	43.3	44.9	13.0
Moscow	42.8	50.1	53.4	9.1
Volgograd	42.7	33.5	37.4	5.7
Average indicator	45.0	38.7	40.7	11.4
Khabarovsk	40.5	31.5	38.4	–3.1
Riga	40.3	42.2	44.9	10.7
Chelyabinsk	35.6	49.3	50.8	14.5
Kiev	33.4	57.2	62.4	7.4
Tallinn	32.9	30.0	30.2	15.3
Gorky	32.6	44.0	48.3	8.3
Average indicator	35.9	42.4	45.8	8.8
Tyumen	31.5	34.0	42.4	– 5.3
Alma Ata	30.6	41.4	44.4	10.7
Kemerovo	29.6	35.0	41.1	0.0
Donetsk	27.8	43.9	45.1	13.0
Leningrad	22.3	45.8	47.2	13.0
Omsk	13.9	30.9	36.6	0.0
Average indicator	25.9	38.5	43.7	5.2
USSR (control)	—	20.5	24.3	—

[a]Effectiveness coefficient = $\left(1 - \frac{P_{c'} : P_{t'}}{P_c : P_t}\right) \times 100$

where: P_t is the morbidity indicator in town for the 1980–1985 period, P_c for the USSR (control), $P_{t'}$ and $P_{c'}$ for 1968–1980.

6. CONCLUSIONS

The results obtained prove the economic efficiency and epidemiological feasibility of the complex of anti-influenza measures, and should stimulate efforts towards further expansion and improvements of these measures which are aimed at protecting people against influenza.

REFERENCES

Alexandrova, G.I., Polezhaev, F.I., Budilovsky, G.N. *et al.* (1984) *Infect. Immun.* **44**, 734–739.

Bashlyaeva, Z.A., Sumarokova, A.A., Nefyodova L.A. *et al.* (1986) *J. Mikrobiol.*, **2**, 49–54 (in Russian).

Bykov, S.E. (1980) In *The Epidemiology and Prophylaxis of Influenza*, pp. 129–134. Leningrad: The All-Union Research Institute of Influenza (In Russian).

Belyakov, V.D., Selivanov, A.A., Ivanov, K.G. and Degtyarev, A.A. (1979) In *The Etiology and Specific Influenza Prophylaxis*, pp. 33–37. Leningrad: The Research Institute of Epidemiology and Microbiology (In Russian).

Dowdle, W., Millar, J.D., Shonberger, L.B. *et al.* (1980) *J. Infect. Dis.*, **141**, 258–264.

Gagarinova, V.M. , Kustikova, Yu.G. and Oleinikova, E.V. (1983) *J. Mikrobiol.*, **4**, 50–53 (In Russian)

Gagarinova, V.M., Kustikova, Yu.G., Oleinikova, E.V. and Shadrin, A.S. (1982) In *The Etiology and Diagnostics of Influenza and Other Febrile Respiratory Diseases*, pp. 32–37. Leningrad: The All-Union Research Institute of Influenza (In Russian).

Gagarinova, V.M., Araslanova, I.I. and Dekterev, A.N. (1983a) *J. Mikrobiol.* **5**, 67–70 (In Russian).

Gagarinova, V.M., Shadrin, A.S., Kubar, O.I. *et al.* (1983b) *J. Mikrobiol.*, **3** 60–63 (In Russian).

Ignatieva, G.S., Maksakova, V.L., Yefremov, B.V. *et al.* (1983) In *The Epidemiology of Influenza and Other Acute Respiratory Illnesses*, pp. 116–121. Leningrad: The All-Union Research Institute of Influenza (In Russian)

Ignatieva, G.S., Shadrin, A.S., Maksakova, V.L. *et al.* (1986). *J. Mikrobiol.*, **3**, 8–10 (In Russian).

Ivanova, S.A. and Gagarinova, V.M. (1981) In *Comprehensive Influenza Prophylaxis*, pp. 24–32. Leningrad: The All-Union Research Institute of Influenza (In Russian).

Karpova, L.S., Zvereva, E.P. and Boldasov, V.K. (1983) In *Epidemiologiya grippa i drugikh ORZ* (The Epidemiology of Influenza and Other Febrile Respiratory Diseases), pp. 78–83. Leningrad: The All-Union Research Institute of Influenza (In Russian).

Karpukhin, G.I. (Ed.) (1986) Gripp (Influenza). Leningrad: Meditsina (In Russian).

Karpukhin, G.I. and Shadrin, A.S. (1977) In *Abstracts of Proceedings of the 16th All-Union Congress of Microbiologists and Epidemiologists, Part I*, pp. 94–98. Moscow: Meditsina (In Russian).

Karpukhin, G.I. (1981) In Comprehensive Influenza Prophylaxis, pp. 9–17. Leningrad: The All-Union Research Institute of Influenza (In Russian).

Karpukhin, G.I. (1983) Economic cooperation of CMEA member-states, **1**, pp.46–47 The All-Union Research Institute of Influenza (In Russian).

Karpukhin, G.N., Shvetsova, E.G. and Malysheva, A.M. (1979) In *Problems of Influenza and Febrile Respiratory Diseases*, pp. 91–121. Leningrad: VNII Grippa (In Russian).

Maksakova, V.L., Ignatieva, G.S., Shadrin, A.S. *et al.* (1985) In *Vaccines and Anti-Influenza Vaccination*, pp. 132–137. Leningrad: The All-Union Research Institute of Influenza (In Russian).

Meiklejohn, G. (1983) *J. Infect. Dis.*, **148**, 755–784

Polezhayev, F.I., Budilovsky, G.N., Garmashova, L.M. *et al.* (1983) *Vopr. Virusol.*, **6**, 710–714 (In Russian).

Romanov, V.L., Zosimenko, I.I. and Baykeyev, Yu.G. (1982) *Voennomedits. J.*, **2**, 34–37 (In Russian)

Shadrin, A.S. and Karpukhin, G.I. (1981) In *Comprehensive Influenza Prophylaxis* pp. 17–24. Leningrad: The All-Union Research Institute of Influenza (In Russian)

Shadrin, A.S., Ignatieva, G.S., Yefremov, B.V. *et al.* (1982) In *Kompleksnaya profilaktika grippa* (Comprehensive Influenza Prophylaxis), pp. 25–31. Leningrad: The All-Union Research Institute of Influenza (In Russian).

Shadrin, A.S., Ignatieva, G.S., Yefremov, B.V. *et al.* (1982) In *The Etiology and Diagnostics of Influenza and Other Febrile Respiratory Diseases*, pp. 25–31. Leningrad: The All-Union Research Institute of Influenza (In Russian).

Shadrin, A.S., Yefremov, B.V. and Kustikova, Yu.G. (1983) In *The Etiology of Influenza and Other Febrile Respiratory Diseases*, pp. 84–88. Leningrad: The All-Union Research Institute of Influenza (In Russian).

Shadrin, A.S., Karpukhin, G.I., Gagarinova, V.M. *et al.* (1985) *Z. Erkrank. Atm. Organ.* **165**, 163–170.

Shadrin, A.S. and Zykov, M.P. (1984) *Problems of Comprehensive Prophylaxis*. Moscow: The All-Union Research Institute of Med. and Med. Tech. Inf. (In Russian).

Shadrin, A.S. and Golubev, D.V. (1980) In *Influenza Epidemiology and Prophylaxis*, pp. 5–15. Leningrad: The All-Union Research Institute of Influenza (In Russian).

Shadrin, A.S. and Karpukhin, G.I. (1981) In *Comprehensive Influenza Prophylaxis*, pp. 17–24. Leningrad: The All-Union Research Institute of Influenza (In Russian).

Shadrin, A.S., Malysheva, A.M., Lugovskaya, V.P. *et al.* (1982) *J. Mikrobiol.*, **9**, 78–81 (In Russian).

Shadrin, A.S., Malysheva, A.M., Lugovskaya, V.P. *et al.* (1984) *Eksp. Klin. Farmakoter. Riga*, **13**, 137–142 (In Russian).

Yerofeyeva, M.N., Shadrin, A.S., Lugovskaya, V.P. *et al.* (1982) In *The Clinical Aspect and Treatment of Influenza and Other Acute Respiratory Illnesses*, pp. 12–19. Leningrad: The All-Union Research Institute of Influenza (In Russian).

Sov. Med. Rev. E. Virology Reviews, Vol. 3. 1990, pp. 131–150
Photocopying permitted by license only

DISTINCTIVE ANTIGENIC DRIFT IN THE INFLUENZA A(H3N2) VIRUS HEMAGGLUTININ

M.A. YAKHNO, N.A. PETROV, S.S. YAMNIKOVA, S.K. VASILENKO, and L.Ya. ZAKSTELSKAYA

D.I. Ivanovskii Institute of Virology, USSR Academy of Sciences, Moscow and All-Union Research Institute of Molecular Biology, Scientific and Production Association 'Vektor', USSR Ministry of the Medical and Microbiological Industry, Novosibirsk

Abstract

Variation in the H3 hemagglutinin (HA) among influenza A(H3N2) strains isolated in the USSR since the emergence of a novel shift variant, A/Hong Kong/1/68, has been comprehensively analysed. It has been found that the antigenic composition of the H3 HA changed during the period of observation due to changes in the primary structure of the gene coding for the HA protein. The progressive heterogeneity of the antigenic composition of the H3 HA in the strain population has been noted.

However, regardless of population heterogeneity, each epidemic wave was associated with distinct leader strains that probably featured a combination of the most favourable mutations in the gene providing for their epidemicity. During the period of observation from 1968 to 1986, all such strains originally arose abroad, and were subsequently introduced into the USSR at various dates.

The appearance of antigenic strains — revertants and relicts — with the HA's higher antigenic affinity for drift variants, including the prototype A/Hong Kong/1/68 strain, has been demonstrated; though previously dominant in active circulation, these variants were ousted from it.

1. INTRODUCTION

In spite of the fact that the variability of influenza virus, in particular that of its protective antigens hemagglutinin (HA) and neuraminidase (NA), has been the subject of intensive study over recent decades, it remains not only uncontrollable and unpredictable but also largely unknown.

Rotating in succession, different antigenic variants of the influenza A(H3N2) virus have been actively circulating in the human population for nearly 20 years now.

A new shift variant of the influenza A virus, A(H3N2), appeared in circulation in 1968 causing a new influenza pandemic. Its HA was totally different from that of the A(H2N2) virus but closely related to that of the strain detected in ducks in 1963 (Zakstelskaya *et al.*, 1969a and b).

The emergence of the new strain with shift variation in the HA caused three successive diffused influenza A waves in the USSR in 1968–1969, 1969–1970 and 1971–1972. The A(H3N2) viruses became dominant in the epidemic process and rapidly, virtually within one season, entirely ousted the A(H2N2) strain from circulation. An estimated nine epidemic waves of influenza A(H3N2) were reported worldwide from 1968 to 1985. The question that arose related to the extent to which this was associated with the change in the agent.

2. ANTIGENIC DRIFT

Our comparative studies on WHO reference strains, isolated in various years by the hemagglutinin-inhibition (HI) test with hyperimmune rat sera, revealed a distinct antigenic drift in the HA (Table 1).

This drift was primarily characterized by a progressive decrease in the amount of antigenic determinants cross-reacting with antibodies to the prototype A/Hong Kong/1/68 strain all the way to their complete loss in the drift variants that appeared in the 1977–1979 and later.

The drift variants of 1982–1986 were indirectly related to A/Hong Kong/1/68 via the A/Port Chalmers/1/73, and A/Victoria/3/75 drift variants. The intensity of shift between the successively arising drift variants varied from year to year. Thus, for instance, the A/Victoria/3/75 strain had only 6.3% homology to A/Port Chalmers/1/73. whereas A/Texas/1/77 and A/Bangkok/1/79 retained as much as 25% of common determinants.

A comparison of the HA of reference drift variants recommended by WHO with strains isolated during the course of influenza epidemics in the USSR, revealed that strains, similar to a drift variant that had occurred abroad, tended to appear in this country several months later

Table 1 Antigenic drift of A (H3N2) viruses 1968–1985.

Antigens	*Antisera*									
	1	*2*	*3*	*4*	*5*	*6*	*7*	*8*	*9*	*10*
1. A/Hong Kong/1/68	100	50	3	1.5	0[a]	0	0	0	0	0
2. A/England/42/72	12.5	100	50	25	12.5	12.5	0	0	6.3	2.5
3. A/Port Chalmers/1/73	3	25	100	6.3	25	25	1.5	25	25	6.3
4. A/Victoria/3/75	12.5	3.1	6.3	100	6.3	12.5	0	0	25	3.1
5. A/Texas/1/77	0.75	25	6.3	12.5	100	25	25	50	35	6.3
6. A/Bangkok/1/79	1.5	25	6.3	25	25	100	25	50	12.5	50
7. A/Belgium/2/81	0.4	0	0.75	25	50	100	100	50	012.5	100
8. A/Philippines/2/82	0	0	1.5	1.5	1.5	12.5	12.5	100	12.5	12.5
9. Caen/1/84	0	0	0.4	12.5	25	6.2	6.2	25	100	25
10. A/Mississippi/1/85	0	0	0	12.5	25	12.5	6.2	100	25	100

[a] Less than 0.3%.
The figures indicate percentage homology of antigenic determinants in the hemagglutinin.

(e.g. A/England/42/72 and A/Bangkok/1/79 drift variants), sometimes a year or a year and a half later (A/Port Chalmers/1/73). The first A/England/42/72 drift variant appeared in the USSR in late 1972–early 1973 so that it restricted the circulation of A/Hong Kong/1/68-like strains and subsequently ousted them altogether.

The A/Port Chalmers/1/73 variant did not arise in the USSR until late in 1974 when it caused, in the winter of 1974–1975, an epidemic wave in which, however, strains similar to the A/England/42/72 and to the slightly epidemic A/Scotland/840/74 variant kept circulating on a limited scale.

The A/Victoria/3/75 variant appeared in the USSR in the fourth quarter of 1975. It became dominant and stimulated the emergence of a new influenza A wave in the 1975–1976 season, with minor involvement of strains related to A/Port Chalmers/1/73.

A new drift variant, A/Philippines/2/82, arose and spread in the season of 1982–1983 and 1983–1984. Strains similar to A/Bangkok/1/79 and A/Texas/1/77 as well as to a low activity variant, A/Shanghai/31/80, went on circulating, albeit in a minor proportion.

The appearance in 1985 of a new drift variant, A/Caen/1/84, in the USSR did not completely replace circulating strains similar to A/Philippines/2/82, A/Bangkok/1/79, and even A/Texas/1/77, and the 1985–1986 season saw the emergence of strains identical with a new drift variant, A/Mississippi/1/85.

Thus, recurrent epidemic waves of A(H3N2) influenza involved the appearance and diffusion of new antigenic drift variants that used to arise outside the USSR in different regions of the world.

Surveillance of circulating influenza A(H3N2) strains revealed an original phenomenon: heterogeneity of a drift variant population with respect to the HA antigenic composition, which varied in intensity from year to year. For example, strains isolated in 1979 in the USSR proved to be highly heterogeneous, as was clearly manifest from their interaction with monoclonal antibodies (MABs) (Table 2) to the prototype A/Hong Kong/1/68 strain and the equine influenza virus A/Eq/Miami/1/63, whose HAs belong to the H3 group. Some of them, such as [A/USSR/127/79,] A/USSR/201/79, and A/USSR/279/79, did not react with MABs to the A/Hong Kong/1/68 and A/Eq/Miami/1/63 viruses. Others (A/USSR/215/79, A/USSR/301/79, and A/USSR/310/79) retained the characteristics of the reference A/Bangkok/1/79 virus — that is, they did not react with serum to equine influenza and were recognized only by MAB 63 to the reference A/Hong Kong/1/68 virus, that recognized all the preceding reference A(H3N2) viruses of 1968–1979 but A/Victoria/3/75.

However, identified among these viruses were strains exhibiting an opposite trend. One of them (A/USSR/157/79) was recognized by all

Table 2 Characterization of influenza A (H3N2) viruses (1979 and 1986) by the HI test with monoclonal antibodies.

Monoclonal antibodies	*A/Eq/Miami/1/63*						*A/Hong Kong/1/68*						*A/HK/1/68 polyclonal antisera*
	1	*2*	*3*	*4*	*5*	*6*	*3*	*19*	*21*	*24*	*63*	*228*	
A/Hong Kong/1/68	800	1600	3200	12800	100	3200	12800	12800	12800	12800	800	1600	320
A/Port Chalmers/1/73	200	1600	1600	<	<	1600	<	800	<	<	12800	1600	80
A/Victoria/3/75	<	<	<	<	<	3200	<	1600	<	<	<	<	40
A/Texas/1/77	200	<	<	<	<	<	<	<	<	<	1600	<	0
A/Bangkok/1/79	<	<	<	<	<	<	<	<	<	<	1600	<	0
A/USSR/127/79	<	<	<	<	<	<	<	<	<	<	<	<	0
A/USSR/201/79	<	<	<	<	<	<	<	<	<	<	<	<	0
A/USSR/279/79	<	<	<	<	<	<	<	<	<	<	<	<	0
A/USSR/215/79	<	<	<	<	<	<	<	<	<	<	3200	<	0
A/USSR/301/79	<	<	<	<	<	<	<	<	<	<	3200	<	0
A/USSR/310/79	<	<	<	<	<	<	<	<	<	<	3200	<	0
A/USSR/157/79	<	<	100	<	100	<	12800	12800	12800	12800	12800	12800	40
A/USSR/174/79	400	400	12800	100	800	100	<	100	<	<	1600	<	0
A/USSR/393/86	200	<	<	<	<	<	12800	3200	12800	12800	12800	12800	320
A/USSR/394/86	800	<	200	<	<	<	12800	400	12800	12800	12800	12800	320

Values are the reciprocals of monoclonal antibody titres; < denotes titres less than 100.

MABs to the A/Hong Kong/1/68 virus used, but without reacting with monoclonals to the equine viruses. In contrast, another strain (A/USSR/174/79) reacted with four out of six monoclonals to the A/Eq/Miami/1/63 virus and with monoclonal 63 to the conservative determinant of the A/Hong Kong/1/68 virus HA.

The tendency toward an increase in the heterogeneity of the influenza A(H3N2) strains continued in the following years. In 1986, the influenza A(H3N2) strains kept circulating in the USSR during an epidemic caused by the influenza A(H1N1) virus. These were mainly Philippine- and Mississippi-like variants, but identified among them were two strains (A/USSR/393/86 and A/USSR/394/86) obtained from 2-year-old children at Kaluga, that would not be neutralized by sera to current reference A(H3N2) viruses.

Studying the uniqueness of the HA of the A/USSR/393/86 and A/USSR/394/86 strains by the HI test with polyclonal sera (Table 3) revealed that these strains were similar to the reference A/Hong Kong/1/68 virus and partially similar to the A/duck/Ukraine/1/63, A/Eq/Miami/1/63, and A/Port Chalmers/1/73 strains. As distinct from the prototype A/Hong Kong/1/68 strain that did not react with the antisera of strains isolated in 1975–1979, the new A/USSR/393/86 and A/USSR/394/86 strains were slightly inhibited (down to 1/16–1/32 of homologous titre) by antisera to these reference viruses. However, A/USSR/393/86, A/USSR/394/86 and A/Hong Kong/1/68 did not react with any serum of the influenza A(H3N2) viruses of 1982–1985.

By contrast, the new strains were distinguished by a broader spectrum in the Hong Kong series, since antisera prepared against them, identified related determinants in the HA of influenza A(H3N2) viruses isolated between 1977 and 1985 or, in other words, they contained antigenic determinants peculiar to these viruses (Table 4).

Analysis of the HA antigenic composition with a panel of MABs to the HA of the A/Hong Kong/1/68 and A/Eq/Miami/1/63 viruses (Table 2) also revealed dissimilarity between the A/Hong Kong/1/68, A/USSR/393/86 and A/USSR/394/86 strains. While being actively neutralized by all of the MABs to the A/Hong Kong/1/68 HA used, they intereacted with MABs to the A/Eq/Miami/1/63 HA in different ways. As opposed to A/Hong Kong/1/68, the A/USSR/393/86 and A/USSR/394/86 strains had already lost epitopes recognized by two, four, five and six MABS to the A/Eq/Miami/1/63 HA. All this indicates that Hong Kong-like strains, A/USSR/393/86 and A/USSR/394/86, are original variants which, while being more closely related to A/Hong Kong/1/68, do have significant distinctions. The cases described are not an exception. Isolation of a Hong Kong-like virus from children in Australia was also reported in 1979 (Moore *et al.*, 1981).

Table 3 HI test of the 1986 Hong Kong-like strains with polycional antibodies to group H3 influenza A viruses.

	Antibodies									
Antigens	*A/Eq/ Miami/63*	*A/duck/ Ukraine/ 1/63*	*A/Hong Kong/1/68*	*A/port Chalmers/ 1/73*	*A/Victoria/ 3/75*	*A/Texas/ 1/77*	*A/Bang-kok/1/79*	*A/Philippi-nes/2/82*	*A/Caen/ 1/84*	*A/Miss/ 1/85*
A/Hong Kong/1/68	160*	320	320	160	<20	<20	<20	<20	<20	<20
A/USSR/393/86	640	320	320	80	<20	40	<20	<20	<20	<20
A/USSR/394/86	320	640	320	80	<20	40	<20	<20	<20	<20
Homologous titre	2560	1280	320	2560	320	1280	320	320	2560	1280

* Reciprocal dilutions of sera completely inhibiting 4–8 agglutination units of the virus.

Table 4 A comparison of the activities of polyclonal sera in the HI test towards new Hong Kong-like strains versus influenza A (H3N2) viruses, 1968–1985.

	Polyclonal sera		
Viruses	*A/Hong Kong/1/68*	*A/USSR/393/86*	*A/USSR/394/86*
A/Hong Kong/1/68	1280	2560	2560
A/England/42/72	320	1280	1280
A/Port Chambers/1/73	160	160	160
A/Victoria/3/75	40	160	320
A/Texas/1/77	0	80	40
A/Bangkok/1/79	0	160	160
A/Philippines/2/82	0	160	160
A/Caen/1/84	0	320	160
A/Mississippi/1/85	0	40	80
A/USSR/393/86	1280	2560	1280
A/USSR/394/86	1280	2560	2560

0 denotes titre less than 20.

3. INVESTIGATION OF THE NATURE OF "OLD" STRAINS

Since the appearance of "old" strains represents an epidemiological reality, a thorough investigation of their nature is in order. With the present state of the art, the most informative approach seems to be one that comprehensively compares the primary genome structures of freshly isolated viral 'anachronisms' with the genes of prototype strains. With this in mind, studies were conducted to elucidate the sequence of nucleotides in the gene of the basic antigenic protein which is the HA of the Hong Kong-like A(H3N2) strains isolated in 1986.

The sequence of bases in the gene was determined through elongating 5′-labelled oligonucleotide primers by reverse transcriptase on the template of the overall virion RNA, followed by sequencing the resulting cDNA (Zhdanov *et al.*, in press). A modified method (Beklemishev *et al.*, 1984) was followed to obtain viral RNA from a viral preparation freshly purified by sucrose density gradient centrifugation (Shilov *et al.*, 1981). Primers corresponding to positive chain regions 15–34, 185–199, 339–355, 487–499, 564–577, 621–637, 789–802, 966–982, 1245–1262 and 1485–1501 of the A/Aichi/2/68 strain HA gene were used (Verhoeyen *et al.*, 1980).

Figure 1 compares the primary HA gene structures of the Hong Kong-like A/USSR/393/86 and A/USSR/394/86 strains and those of the reference pandemic A/Hong Kong/1/68 strain (Gething *et al.*, 1980).

	(A) 168	260	550	1076	1169	1511	(B) HA1: 31	158	HA2: 149
A/Hong Kong/1/68	G	A	G	C	A	A	D	G	I
A/USSR/393/86	A	T	A	T	A	G	N	E	M
A/USSR/394/86	A	T	A	T	C	G	N	E	M

Figure 1 (A) Distinctions between the nucleotide sequence of the HA genes in the current Hong Kong-like influenza A(H3N2) viruses and that of the A/Hong Kong/1/68 gene. Residues are numbered from the 5′-terminus of the positive chain. (B). Corresponding changes in the amino acid sequence of the protein. Numbering proceeds from the N-end of the appropriate fragments of the mature HA molecule.

A review of our findings in comparison with the evidence presented in the literature leads to several inferences. On the one hand, the 1986 Hong Kong-like strains differed fairly little in genome structure from one another, as well as from the 1968 reference strains (five to six substitutions). For comparison, it should be noted that during the course of antigenic drift from 1968 to 1985 over 100 nucleotide mutations accumulated in the HA gene of the viruses of this serosubtype (Petrov *et al.*, 1986 and Both *et al.*, 1983). Significantly, each new drift variant's population contained both mutations leading to amino acid substitution and 'silent' nucleotide substitutions in the HA gene, which unequivocally established their relatedness to subsequent isolates. Evidence on the HA gene primary structure of A/USSR/393/86 and A/USSR/394/86 demonstrates unambiguously that they are genetically closely related to A/Hong Kong/1/68 and do not naturally result from the processes of antigenic drift of preceding strains.

On the other hand, the differences revealed preclude identification of the gene coding for the HA of new Hong Kong-like viruses with the HA gene of the laboratory prototype A/Hong Kong/1/68 strain and of other previously described influenza virus A(H3N2) strains.

The nature and arrangement of the distinctions recognized in the amino acid sequence of the 1968 strains (Figure 1B) show that they are associated with changes in the receptor properties of these viruses, rather than with antigenic drift, i.e. with the selection of mutants in response to antibodies in a partially immune human population. Such selection is known to result in the concentration of the bulk of the mutations in circumscribed portions of the molecule, which are the antigenic determinants A, B, C and D (Petrov *et al.*, 1986; Wilson *et al.*, 1981). In the strains of 1986, one of the two substitutions identified in the HA1 subunit (158 Gly→Glu) is located in the antigenic determinant B

(Wilson *et al.*, 1981), and might be responsible for the changes in the virion's antigenic characteristics noted above. Nevertheless, it is known that this mutation may also serve an alternative function concerned with variation in tissue tropism (Both *et al.*, 1983). The remaining two substitutions of the amino acid residues; 31 Asp→Asn in HA1 and 149 Ile→ Met in HA2, are conservative in character, widely separated from the HA molecule functional sites, and probably neutral.

The observed amino acid and nucleotide distinctions all favour authenticity of the Hong Kong-like viruses studied. An equally important argument is also provided by the reassortative nature of the A/USSR/394/86 virus genome with respect to gene M with the current influenza viruses of the A(H1N1) serosubtype (Sandakhchiyev *et al.*, in press).

There are four possible explanations for this phenomenon:

1. laboratory strains may have been released into the natural environment;
2. Hong Kong-like strains may have been introduced into the human population from avian reservoirs;
3. the basically unexplored possibility of the long-term preservation of the genome in its 'entirety' virtually without any gene mutations and reassortment in the form of a 'dormant' virus with subsequent reactivation;
4. a distinctive direction that may have been taken by the accumulated mutations in the HA gene at the later stages of antigenic drift, or reversion accumulation, the examples of which will be discussed below.

A more detailed examination of the remaining seven Hong Kong-like isolates (1986) and several other similar viruses, as reported by Sandakhchiyev *et al.* (in press), leads to the excluding of two out of the four possible explanations for the 'relict' strain phenomenon presented above: it cannot be accounted for by the introduction into the human population of virus genes from avian reservoirs nor by reversion accumulation during antigenic drift. Nor is it attributable in this case to the repeated transfer into the natural circulation of quite different viruses (Sandakhchiyev *et al.*, in press) distinguishable from the reference strains produced in the laboratory. The most plausible explanation may come, in our opinion, from the truly relict nature of these viruses. Further study is now needed to elucidate the specific mechanisms by which human influenza A virus genomes, that appear now and again as antigenic anachronisms, can be maintained for decades, virtually without mutation.

4. THE EXTENT OF HETEROGENEITY IN THE ANTIGENIC DRIFT IN INFLUENZA A(H3N2) HA

In order to define the extent of heterogeneity in the course of antigenic drift in the influenza A(H3N2) virus HA, we examined 230 epidemic strains of this virus isolated in the USSR during the 1985 epidemic season. Antigenic analysis of the HA of these strains, by the HI test with polyclonal sera, revealed their marked group heterogeneity and indicated that with respect to HA specificity they were different from the 1982 reference A/Philippines/2/82 strain, showing a more pronounced homology to the 1984 reference A/Caen/1/84 strain. Also noteworthy was the fact that some of the epidemic strains, which were closely related to the reference A/Caen/1/84 strain, had a marked affinity for the 1977 reference A/Texas/1/77 strain and were neutralized by its serum up to 1/2–1 of the homologous titre.

As seen from Table 5, the 1985 strains fall into three groups by degree of antigenic relationship to the reference variants of previous years:

1. strains most actively interacting with the reference A/Philippines/2/82 antiserum;
2. strains related to the reference A/Caen/1/84 strain;
3. a representative of the influenza A(H3N2) virus strains circulating in late 1985–early 1986, A/USSR/281/86, completely neutralized by the A/Mississippi/1/85 antiserum.

It is noteworthy that the strains of the second and third groups, along with pronounced affinity for the 1985 and 1986 reference strains, were active, up to 1/2–1/4 of the homologous titre, in reacting with the reference A/Texas/1/77 antiserum. Thus, the 1985 influenza A(H3N2) virus epidemic strains represented a heterogeneous group including several antigenic variants, incorporating strains that have been in circulation since 1977. Elucidating the nature of the strains of groups 1 and 2 was of interest and merited attention.

In order to gain a deeper insight into the HA antigenic structure and refine the results obtained, we utilized a panel of monoclonal antibodies to the HA of the 1977 and 1982 reference strains, kindly presented to us by Dr R. Webster, USA, and Dr J. Skehel, Great Britain.

The strains studied interacted with monoclonal antibodies in different ways (Table 6). For example, the strains of group 1 were identified by 12 out of the 13 used monoclonal antibodies to the A/Philippines/2/82 strain HA without essentially interacting with monclonal antibodies to the A/Texas/1/77 HA. Each strain of group 2 had already lost four to five epitopes recognized by monoclonal antibodies to A/Philippines/2/82. At the same time, the strains of this group, as well as the reference A/

Table 5 Antigenic characterization of the HA of the 1985 influenza A(H3N2) viruses by the HI test.

		Sera				
Antigens	*Groups*	*A/Texas/1/77*	*A/Bangkok/1/79*	*A/Philippines/2/82*	*A/Caen/1/84*	*A/Mississippi/1/85*
A/Texas/1/77		*640*[a]	160	640	640	160
A/Bangkok/1/79		80	640	1280	160	1280
A/Philippines/2/82	I	20	80	*1280*	80	160
A/USSR/3/85		80	160	1280	160	1280
A/USSR/8/85		80	160	1280	320	320
A/USSR/26/85		40	160	640	320	1280
A/Caen/1/84	II	320	80	160	*640*	160
A/USSR/2/85		160	80	80	320	160
A/USSR/31/85		320	40	160	640	320
A/Mississippi/1/85	III	160	160	640	320	1280
A/USSR/281/86		320	640	640	160	1280

[a]Reciprocals of polyclonal serum titres.

Table 6 Antigenic analysis of the HA of influenza A(H3N2) versus (1985–1986) with monoclonal antibodies in the HI test.

	A/Texas/1/77 monoclonal antibodies						*A/Philippines/2/82 monoclonal antibodies*												
Viruses	29	39	52	77	59	185	16	20	13	29	10	19	3	25	30	12	14	5	11
A/Philippines/2/82	1600[a]	<	200	<	<	<	6400	800	1600	12800	1600	400	12800	6400	6400	6400	12800	6400	12800
A/USSR/3/85	320	<	<	<	<	<	<	800	6400	12800	800	1600	12800	12800	12800	6400	12800	6400	12800
A/USSR/8/85	160	<	200	200	<	200	<	800	800	800	1600	800	6400	6400	12800	6400	6400	6400	12800
A/USSR/26/85	<	<	<	<	<	400	<	400	400	12800	800	800	6400	6400	3200	3200	6400	6400	12800
A/Texas/1/77	6400	6400	12800	12800	12800	12800	<	<	<	<	1600	200	12800	6400	12800	6400	12800	12800	12800
A/Caen/1/84	3200	<	3200	6400	6400	25600	<	<	<	<	800	400	12800	12800	12800	6400	12800	12800	12800
A/USSR/2/85	6400	<	12800	3200	12800	51200	<	<	<	<	<	800	6400	12800	12800	6400	12800	6400	6400
A/USSR/31/85	3200	<	1600	3200	3200	6400	<	<	<	<	400	400	12800	6400	6400	6400	12800	12800	12800
A/Mississippi/1/85	800	<	400	400	<	400	12800	3200	1600	12800	3200	3200	12800	12800	12800	12800	12800	12800	12800
A/USSR/281/86	6400	<	1600	1600	<	200	12800	1600	3200	12800	6400	6400	12800	12800	12800	12800	12800	12800	12800

[a]Values are to the reciprocals of monoclonal antibody titres;
< denotes titres less than 200.

Caen/1/84 strain, interacted vigorously with five out of the six used monoclonal antibodies to the A/Texas/1/77 HA. The A/Mississippi/1/85 and A/USSR/281/86 strains of the third group reacted actively with monoclonal antibodies to the HA of both A/Texas/1/77 and A/Philippines/2/82.

Thus, analysis of the HA antigenic structure using MABs confirmed pronounced antigenic heterogeneity, revealing that, unlike the influenza A(H3N2) viruses circulating in the USSR in 1983 (Zhdanov *et al.*, 1969), epitopes recognizable by MABs to the A/Texas/1/77 HA might have been regained by the 1985 and 1986 strains in the course of drift. To elucidate the mechanism of this phenomenon in the 1985 influenza A(H3N2) virus strains, we unravelled the HA primary structure in representatives of each of the three antigenic subgroups — Philippine-, Caen-, and Mississippi-like (A/USSR/3/85, A/USSR/2/85 [and A/USSR/281/85], respectively (Zhdanov *et al.*, in press).

Comparative analysis of nucleotide sequences revealed that the HA genes of the three strains are closely related. Their common predecessor is most closely related to the Philippine-like A/Netherlands/246/82 strain (Petrov *et al.*, 1986; Skehel *et al.*, 1983). This is supported, among other things, by the fact that amino acid substitutions typical only for the A/Netherlands/246/82 strain and lacking in other influenza viruses (1977–1982) have been detected in all the three viruses at positions 173, 248 and 307 from the N-terminal of the heavy HA subunit. A late representative of the Philippine-like A/USSR/3/85 viruses was closer in structure to the Caen/84 subgroup viruses (A/USSR/2/85 than to the Mississippi/85 subgroup (A/USSR/281/86). The latter must have branched off the main course of natural drift even before 1984, evolving independently since then (Petrov *et al.*, 1986). This indicates that of late distinct variants of the Philippine subgroup have coexisted for a fairly extended time, which is thought to have markedly increased the composition heterogeneity of current viruses.

A comparison of the amino acid sequences of the heavy subunits revealed that all the three viruses studied differed from the 1980–1982 strains in 146 Gly → Ser, a substitution in the domain of antigenic determinant A (Figure 2) (Webster and Laver, 1980). This mutation restores the amino acid sequence inherent in the agents of preceding epidemics — A/Bangkok/1/79 and A/England/321/77. Strains circulating prior to 1982 also displayed the trend toward fixing reversions to earlier variants at later stages of the H3N2 subtype epidemic cycle (Petrov *et al.*, 1986).

Members of the Philippine- and Caen-like subgroups (A/USSR/3/85 and A/USSR/2/85) show a mutation, 144 Asp →Val, previously not encountered in the antigenic determinant and absent in earlier Philippine-like A/Hong Kong/1/82 and Mississippi-like A/USSR/281/

	2	48	*50*	*53*	*54*	57	62	63	83	92	122
1. Queensland/7/70	D	T	K	N	N	I	I	D	T	K	T
2. Texas/1/77	N	T	R	D	S	I	K	N	K	K	N
3. England/321/77	N	T	R	D	S	I	K	N	K	K	N
4. Bangkok/2/79	N	T	R	D	S	I	K	N	K	K	N
5. Hong Kong/1/82	N	A	R	D	S	I	K	N	K	K	N
6. Netherlands/246/82	N	T	R	D	S	V	K	N	K	R	N
7. USSR/281/86	K	T	R	D	S	I	K	N	K	K	N
8. USSR/2/85	K	T	R	D	S	I	K	N	K	K	N
9. USSR/3/85	K	T	R	D	S	I	K	N	K	K	N

	126	128	*129*	*133*	*137*	*143*	*144*	*145*	*146*	*155*	*156*	*158*	*160*	*163*	*170*
1.	T	T	E	Q	N	P	G	S	G	T	K	G	T	V	N
2.	N	T	G	Q	Y	P	D	N	G	Y	K	E	T	V	N
3.	N	T	G	Q	Y	P	D	N	S	Y	K	E	T	V	N
4.	N	T	G	S	Y	S	D	N	S	Y	E	E	K	V	N
5.	N	T	G	S	Y	S	G	N	G	Y	E	E	K	A	N
6.	N	T	G	S	Y	S	D	N	G	Y	E	E	K	V	D
7.	N	A	G	S	Y	S	D	N	S	Y	E	E	K	V	N
8.	N	T	G	S	Y	S	V	N	S	Y	K	E	K	A	N
9.	N	T	G	S	Y	S	V	N	S	Y	E	E	K	V	N

	172	*173*	180	*188*	*189*	*193*	*197*	201	207	213	217	226	242	244	246
1.	D	N	G	N	Q	S	Q	R	R	I	I	Q	V	V	N
2.	G	N	G	D	K	N	Q	R	K	I	V	L	I	L	N
3.	G	N	G	D	K	N	Q	R	K	I	V	L	I	L	N
4.	G	N	G	D	K	K	R	R	K	I	I	L	I	L	N
5.	G	N	G	D	K	N	R	R	K	V	I	L	I	L	N
6.	G	K	I	D	K	N	R	R	K	V	I	L	I	L	N
7.	G	K	G	D	K	N	R	R	K	V	I	Q	I	L	H
8.	G	K	G	D	K	N	R	R	K	V	I	Q	I	L	N
9.	G	K	G	D	K	N	R	K	K	V	I	L	I	L	N

	248	260	*275*	*278*	279	307	323
1.	N	M	D	I	S	K	V
2.	N	I	G	S	S	K	V
3.	N	I	G	S	S	K	V
4.	N	I	G	S	S	K	V
5.	S	I	G	S	S	K	I
6.	T	I	G	S	S	R	V
7.	T	I	G	S	S	R	V
8.	T	I	G	S	S	R	V
9.	T	I	G	S	F	R	V

Figure 2 A comparison of the amino acid sequences in the heavy HA subunits of the epidemic human influenza A(H3N2) viruses replacing successively one another. Amino acid residues are shown only at those positions where mutations became established during the epidemic cycle of this subtype viruses. They are numbered from the N-end of the mature molecule. Underscored are positions situated in the regions of antigenic determinants. For complete HA structures see Sleigh *et al.* (1981) for Queensland/7/70; Both *et al.* (1983) for Texas/1/77 and Bangkok/2/79; Hauptman *et al.* (1983) for England/321/77, and Skehel *et al.* (1983) for Hong Kong/1/82 and Netherlands/246/82.

86 strains. The principal distinction of the A/USSR/2/85 strain of the Caen-like group is due to a mutation, 156 Glu → Lys, in antigenic determinant B. Such a replacement represents a reversion to an amino acid sequence characteristic of the reference A/Texas/1/77 strain and its preceding viruses, which apparently accounts for such a close similarity between the antigenic characteristics of these widely spaced subgroups (Yakhno *et al.*, in press).

The A/USSR/281/86 virus of the Mississippi subgroup differs from all the studied viruses by two mutations (Figure 2). The first mutation , 128 Thr → Ala, is located in the immediate vicinity of the antigenic determinant A, causing the disappearance of the glycosylation site at position 126, that became established in the course of antigenic drift after 1973. It is more than likely that this event could not but have affected the conformation of the chain portion involved in the formation of determinant A leading to a change in the efficacy of binding antibodies to this region of the HA molecule. The second replacement 246 Asn → His, too, causes the glycosylation site to vanish in an area where the HA trimer subunits make contact near antigenic domain D. This site became established after 1980 among viruses of the A/Shanghai/80-A/Hong Kong/82-A/Philippines/82 group via mutation 248 Asn → Ser, whereas its establishment in the drift main branch has been due to 248 Asn → Thr (Figure 2). It is possible that the presence of the site favours a more intimate contact between the subunits in the HA spike (Wilson *et al.*, 1981). [Sterically it contacts a similar glycosylation site at position 165 (Figure 3) (Wilson *et al.*, 1981)]. A three-dimensional oligosaccharide unit located here protects a major part of determinant D against direct contact with antibodies. The occurrence of the same unit at position 246 may result in conformational change. Its disappearance in viruses of the Mississippi branch is bound to restore conformation of the oligosaccharide chain at position 165 and the entire antigenic region D, typical of the A/Bangkok/1/79 viruses. It is this that may account for the similarities between the antigenic characteristics of these two virus groups (Yakhno *et al.*, in press). As far as other detected differences are concerned, the unique 201 Arg → Lysy substitution of the A/USSR/3/85 virus is situated in a region inaccessible for immediate contact with antibodies, in a spacing between the HA trimer subunits. The same, to all appearances, is true of the 279 Ser → Phe mutation, since this amino acid residue faces inwards, the molecule and the conformation of the whole region is rigidly fixed by cysteine bridges 277–68 and 281–305.

Of considerable interest is the 226 Leu → Gln mutation where the HA binds the sialic acid of the cellular receptor in the A/USSR/2/85 and A/USSR/281/86 viruses (Wiley *et al.*, 1981). The mutation restores the structure characteristic of the old A/Queensland/70 influenza virus (Wiley *et al*, 1981). Some studies (Sleigh *et al.*, 1981; Rodgers *et al.*, 1983) have shown that the amino acid residue in question is directly involved in

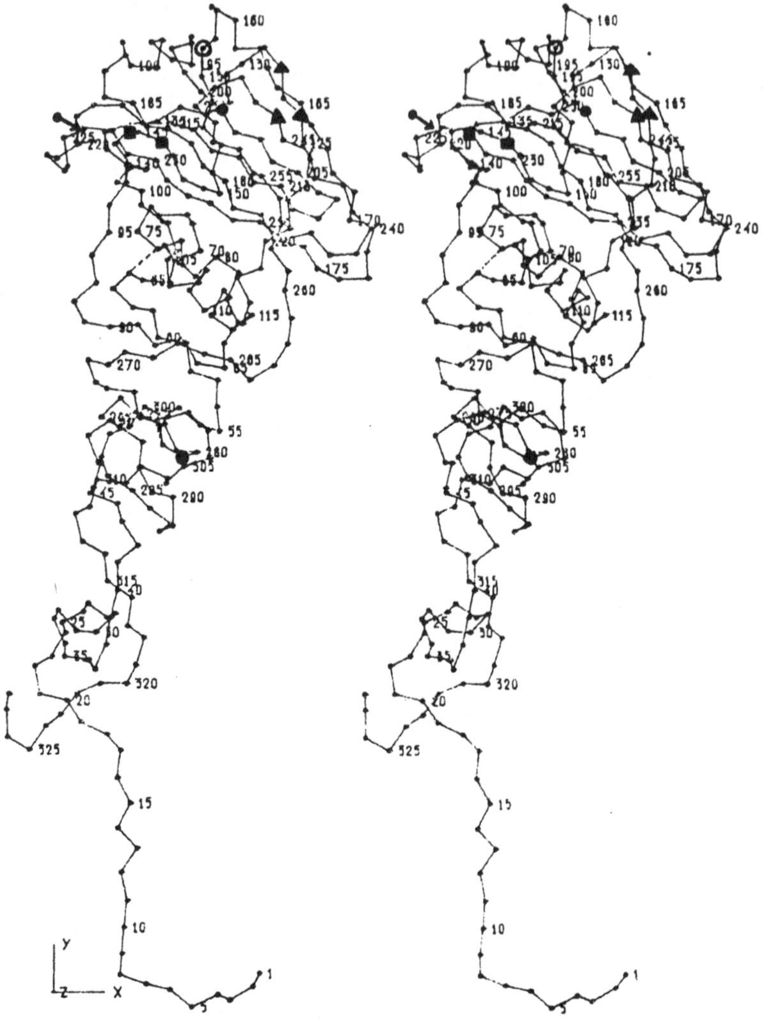

Figure 3 A model diagram (stereogram) illustrating the protein chain folding of the HA molecule heavy subunit of the influenza A(H3) virus (Wilson *et al.*, 1981).
Symbols: ■, mutation 146 Gly→Ser in A/USSR/2/85, A/USSR/3/85 and A/USSR/281/86 viruses; mutation 144 Asp→Val in A/USSR/2/85 and A/USSR/3/85 strains; ⊙, reversion 156 Glu→Lys in A/USSR/2/85 virus; ▲, mutations 246 Asn→His and 128 Thr→Ala resulting in the disappearance of glycosylation sites at positions 126 and 246 in A/USSR/281/86 virus; ●, relatively neutral substitutions 201 Arg→Lys and 279 Ser→Phe in A/USSR/3/85 strain; ↘, mutation 226 Leu→Gln in the recetor zone of the HA of A/USSR/2/85 and A/USSR/281/86 viruses.

recognizing the type of linkage between sialic acid and the preceding sugar residue. Also, the 226 Leu → Gln mutation expands the spectrum of virus-recognizable receptors, decreases the sensitivity to serum γ-inhibitors, and might appropriately broaden the spectrum of permissive tissues.

5. SUMMARY

Our comprehensive antigenic and structural analysis of the HA in the influenza A(H3N2) virus strains isolated in the USSR over 18 years, represents an attempt at systematic research into the antigenic variation of influenza virus within the limits of present molecular and biological methodology. This approach has made it possible to follow, at the molecular level, the time course of changes in the HA antigenic composition during the circulation of the A(H3N2) serosubtype and partially to clarify the mechanisms of progressive heterogeneity of the circulating viruses' antigenic structure. The appearance of both antigenic revertants and 'relict' virus strains has been proved, with the HA being far more closely related to the drift variants, the prototype A/Hong Kong/1/68 included, which previously had been dominant in, but were later ousted from active circulation.

REFERENCES

Beklemishev, A.B., Blinov, V.M., Vasilenko, S.K. *et al.* (1984) *Biorg. Khim* **10**, 1535–1543 (In Russian).

Both, G.W., Sleigh, M.J., Cox, M.J. and Kendal, A.P. (1983) *Virology*, **48**, 52–60.

Both, G.W., Shi, C.H. and Kilbourne, E.D. (1983) *Proc. Natl. Acad. Sci. USA*, **80**, 6996–7000.

Gething, M.J., Bye, J., Skehel, J. and Waterfield, M. (1980) *Nature* (London), **287**, 301–306.

Hauptman, R., Clarkel, D., Mounford, R.G. *et al.* (1983) *J. Gen. Virol*, **64**, 215–220.

Moore, B.W., Webster, R.G., Bean, W.J. *et al.* (1981) *Virology*, 109, 219–220.

Petrov, N.A., Netesov, S.V., Blinov, V.M. and Vasilenko, S.K. (1986) *Mol. Genet. Mikrobiol. Virusol.*, **11**, 7–14 (In Russian).

Rodgers, G.N., Paulson, J.C., Daniels, R.S. *et al.* (1983) *Nature* (London), **304**, 76–78.

Sandakhchiyev, L.S., Petrov, N.A., Yakhno, M.A. *et al. Dokl. Akad. Nauk SSSR* (In press) (In Russian).

Shilov, A.A., Kozlov, Yu.V., Kurmanova, A.G. *et al.* (1981) *Mol. Biol.*, **15**, 1317–1384 (In Russian).

Sleigh, M.J., Both, G.W., Underwood, P.A. *et al.* (1981) *Virology*, **37**, 845–853.

Skehel, J.J., Daniels, R.S., Douglas, A.R. *et al.* (1983) *Bull. WHO*, **61**, 671–676.

Verhoeyen, M.R., Fang, W., Minjou *et al.* (1980) *Nature* (London), **286**, 771–776.

Webster, R.G. and Laver, W.G. (1980) *Virology*, **104**, 139–148.

Wiley, D.C., Wilson, J.A. and Skehel, J.J. (1981) *Nature* (London), **289**, 373–378.

Wilson, L.A., Skehel, J.J. and Wiley, D.C. (1981) *Nature* (London), **289**, 366–373.

Yakhno, M.A., Isachenko, V.A., Yamnikova, S.S. *et al. Vopr. Virusol.* (In press) (In Russian).

Zakstelskaya, L.J., Antonova, I.V., Yevstigneyeva, N.A. *et al.* (1969a) *Vopr. Virusol.* **14**, 320–326 (In Russian).

Zakstelskaya, L.J., Isachenko, V.A., Shenderovich, S.Ph. *et al.* (1969b) *J. Epidemiol.*, **90**, 400–405.

Zhdanov, V.M., Petrov, N.A., Grinev, A.A. *et al. Vopr. Virusol.* (In press) (In Russian).

Zhdanov, V.M., Zakstelskaya, L.J., Isachenko, V.A. *et al.* (1969) *Etiologiya i epidemicheskii protsess pri grippe v sovremennykh usloviyakh* (The Etiology and Epidemiological Process in Influenza Today), pp. 12–18. Leningrad: Research Institute of Influenza, USSR Ministry of Health (In Russian).

Sov. Med. Rev. E. Virology Reviews, Vol. 3. 1990, pp. 151–175
Photocopying permitted by license only

RADIOIMMUNOASSAY OF THE INFLUENZA VIRUS NUCLEOPROTEIN

Ye.N. PROKUDINA and N.P. SEMYONOVA

D.I. Ivanovskii Institute of Virology, USSR Academy of Medical Sciences

Abstract

Presented here are the results of analysing the nucleoprotein (NP) of influenza virus by radioimmunoassay (RIA) on polyethylene film, with the use of monoclonal antibodies (MABs). The method has been shown to provide an effective key to assessing NP variability in influenza A viruses. Exposure of the virus to certain physical agents has led to the assumption that the structure of NP antigenic determinants is fairly complex and that their configuration and arrangement in the NP are dissimilar. The influence of the disintegration of tritium, incorporated into viral DNA, on the capacity of NP antigenic determinants to bind with MABs has been investigated.

The absence of qualitative antigenic variations in different ribonucleoproteins (RNPs) of "standard" and "incomplete" viruses alike has been revealed within a single influenza virus strain. Also, the intracellular NP has been shown to have extracellular virus antigenic determinants during maturation of the influenza virus.

1. INTRODUCTION

Until relatively recently, the internal proteins of the influenza virus had been considered antigenically conservative. (Schild and Pereira, 1969). Of late there have, however, been several reports on the antigenic variability of both the membrane (M) protein (Lecomte and Oxford, 1981) and the nucleoprotein (NP) (Trushinskanya *et al.*, 1985; Van Wyke *et al.*, 1980) of the influenza virus. Specifically, Van Wyke and associates , using a panel of anti-NP monoclonal antibodies (MABs), have detected six groups of variable reactivity among human influenza virus strains and two groups among strains of other animal species. Trushinskaya *et al.* (1985) have related the time of release of human influenza viruses to the capacity of their NP to bind with certain anti-NP MABs. The oligopeptide mapping of the NP has allowed the classification of influenza viruses into several groups (Graisler *et al.*, 1986; Dimmock *et al.*, 1980; Sokolov & Zhdanov, 1982).

Hybridization analysis of the NP gene has brought out the relationship between the species identity of the influenza viruses and the genetic homology of their NP (Bean, 1984; Sholtissek *et al.*, 1985, 1987). It thus appears that influenza virus NP antigenicity is associated with various biological factors and, in addition, there is reason to presume that influenza virus NP is a fairly highly organized protein (Sokolov and Zhdanov, 1982); its antigenicity may to a large extent depend upon the conformation of antigenic determinants and, therefore, upon physical factors.

This review presents evidence for the capacity of the influenza virus NP to bind with MABs, and examines the relationship between this capacity and conformation of the NP on exposure to physical factors, when within different viral ribonucleoproteins (RNPs), as well as during the process of viral reproduction. The effect of β-decay of tritium incorporated into the viral RNA on the antigenicity of the viral NP has been demonstrated. We have used radioimmunoassay (RIA) using polyethylene film.

2. MATERIALS AND METHODS

Influenza viruses: A/WSN/33 (H1N1); A/WS/33 (H1N1); A/PR/8/34 (H1N1); A/PR/8/34 (H1N1) passaged in the lungs of white mice; A/swine/Iowa/33 (H1N1); A/USSR/090/77 (H1N1); A/New Jersey/8/76 (HswN1); A/Aichi/2/68 (H3N2); A/Hong Kong/1/68 (H3N2); A/swine/Hong Kong/3/76 (H3N2); A/swine/Hong Kong/4/76 (H3N2); A/swine/Hong Kong/6/76 (H3N2); A/black-headed gull/Astrakhan/66/82 (H13N2); A/gull/Kazakhstan/470/79 (H1N1).

To produce a "standard" virus, A/WSN/33 (H1N1), embryos were infected at 10 ID_{50}/embryo. "Incomplete" virus was derived from a series of undiluted passages. The A/WSN/33 (H1N1) virus was obtained from the Biophysics Laboratory, and the remaining viruses from the Laboratory of Virus Ecology, Institute of Virology, USSR Academy of Medical Sciences.

Labelling, purification and concentration of the A/WSN/33 (H1N1) virus. The virus was labelled in chick embryos by injecting ^{3}H-uridine (25 Ci mmol $^{-1}$) at 500 − μCi per embryo in 3 h. Purification and concentration of the virus involved first centrifugation, carried out across a 15–60% stepwise sucrose gradient in an SW 27 rotor at 24 000 rpm for 1 h, and then pelleting of the virus suspended in STE under similar centrifugation conditions.

Separation of viral ribonucleoproteins (RNPs) was performed as described by Compans *et al.* (1972). The A/WSN/33 (H1N1) virus, concentrated upon purification in a stepwise sucrose gradient, was suspended in a hypotonic solution with 7% sodium deoxycholate (SDC), layered onto a linear 15–30% sucrose gradient with 0.05% SDC, and centrifuged in an SW 27 rotor at 15 000 rpm for 16 h. Material from gradient fractions was assayed for radioactivity and capacity to bind with anti-Np monoclonal antibodies.

Monoclonal antibodies (MABs) 3/1, 5/1, and 469/6 to the NP of A/WSN/33 (H1N2) were kindly provided by Dr Webster (Van Wyke *et al.*, 1980). MABs H-12 and F-81 to the NP of the A/gull/Kazakhstan/470/79 (H1N1) influenza virus were obtained and kindly supplied by Yamnikova and co-workers (Lvov *et al.* 1986). The MABs were iodinated by the method of Hunter *et al.* (1962). All of the iodinated MABs showed the same specific activity of 10^{8} disintegrations min^{-1}/(μg protein)$^{-1}$.

Radioimmunoassay (RIA). An aliquot of 20μL of the viral preparations under examination was applied to polyethylene film (Prokudina *et al.*, 1986). After adsorption for 5 h, the washed film was immersed in 1% bovine serum albumin (BSA), washed, dried and 10μL of 0.02μCi ^{125}I-MABs were added. Following 1 h incubation, the polyethylene film was washed, dried and autoradiographed. Once the autoradiographs were taken, the film was cut into sections and radiometered in a Packard gamma spectrometer. Samples were considered positive if their radioactivity was more than twice that of a negative control where allantoic fluid from uninfected chick embryos was used instead of the virus.

Applying physical and chemical agents. For ultraviolet (UV) irradiation, polyethylene films overlaid with drops of the A/WSN/33 (H1N1) virus were placed 20 cm beneath an ultraviolet lamp. At variable intervals the films were removed from under the lamp to be treated for further analysis by the RIA technique.

The virus was heated in test tubes placed in a water bath warmed up to 100°C. At variable intervals the tubes were transferred into an ice bath. Further the virus was applied to the film for RIA.

To gain an insight into the effects of different pH on NP antigenicity, the virus was diluted in a citrate-phosphate buffer at pH 7.0, 6.0, 5.0, 4.0 and 3.0, and adsorbed onto the polyethylene film with a view to performing RIA.

3. ASSAYING INFLUENZA VIRUS NP FOR CAPACITY TO BIND WITH MONOCLONAL ANTIBODIES, EMPLOYING RIA ON POLYETHYLENE FILM

Radioimmunoassay results are generally measured in terms of radioactivity counted by gamma spectrometers. However, if preparations are to be assessed qualitatively, an RIA modification with autoradiographic registration of results offers specific advantages in that it is easy to interpret, demonstrate and document, its other merit being an ability to reveal artefacts (Prokudina *et al.*, 1986; Semyonova *et al.*, 1987). This modification subsequently counts radioactivity in samples and shows a sensitivity similar to that of standard RIA.

Type A influenza viruses have been examined for their capacity to interact with anti-NP MABs by RIA with autoradiographic recording of results. The pattern of interaction between the MABs and the viruses in each of the revealed seven groups is shown in an autoradiograph (Figure 1). To obtain this autoradiograph, the viruses (one virus from each group) were allowed to adsorb onto polyethylene film in horizontal rows, while the ^{125}I-MABs were added as vertical rows (two parallels per virus). Dark autographs were taken as indicating the occurrence of binding of the virus to the ^{125}I-MABs (radioactivity being more than three times that in the control); slightly shadowed autographs signified poor binding (radioactivity being 2–2.5 times greater than in the negative control); and no autograph implied no binding.

The autoradiograph vividly demonstrates separation of viruses into groups by their binding with MABs, the members of these groups being designated on the left. The diversity of interaction patterns involving the viruses studied and MABs is supported by evidence of antigenic variability of the influenza virus NP, as reported by several investigators (Trushinskaya *et al.*, 1985; Van Wyke *et al.*, 1980). The classification

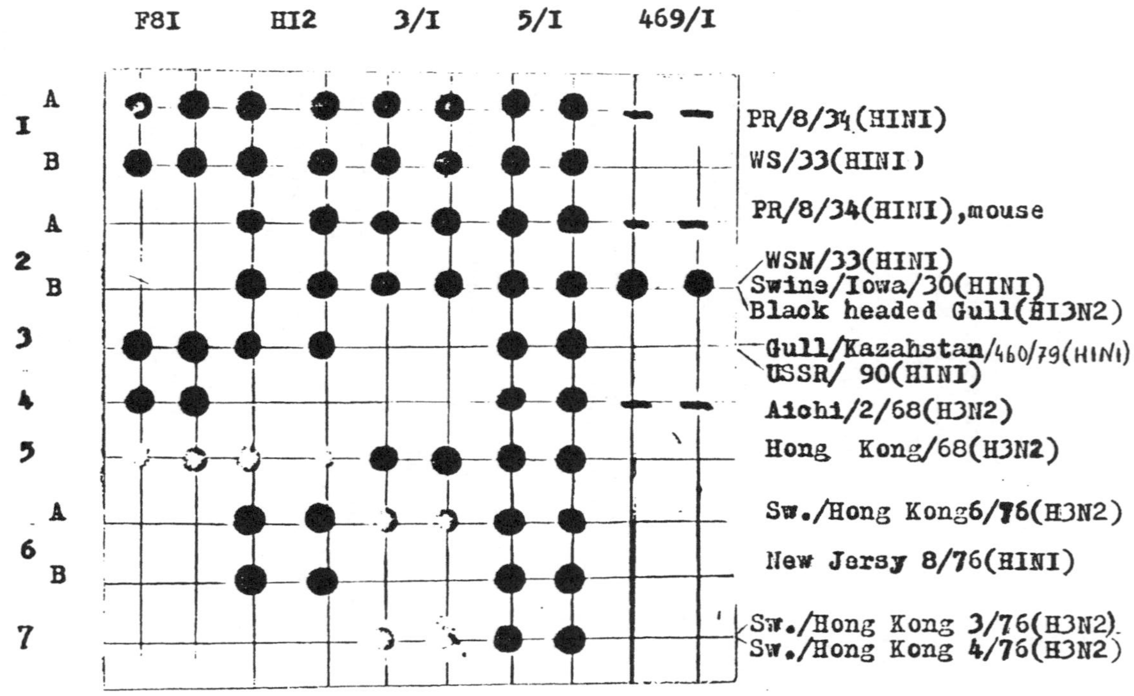

Figure 1 Autoradiograph of results from RIA on film, showing interaction of different influenza A virus strains with anti-NP ^{125}I-MABs. On the right, designations of influenza virus strains (horizontal rows). At the top, designations of anti-NP ^{125}I-MABs (vertical rows, two parallels each). On the left, figures refer to group numbers; letters stand for subgroups. Black autographs, radioactivity is more than five times that of the control. Light authographs, radioactivity is two to three times that of the control. Absent autographs, absent binding;—no reaction performed.

pattern of the viruses does not suggest any correlation between antigenic identity of the nucleoprotein and that of the glycoproteins. For instance, viruses with a similar antigenic characteristic of glycoproteins fell into groups that varied with respect to the antigenic characteristic of nucleoproteins: namely, H1N1 – groups 1, 2 and 3; or H3N2 – groups 4, 5, 6 and 7.

The MABs utilized in this experiment have revealed four antigenic sites in the nucleoprotein of the influenza virus (Semyonova *et al.*, 1987). A somewhat different panel of MABs has brought out five antigenic sites in the NP (Trushinskaya *et al.*, 1985). It should be borne in mind that in the present work we assayed the MABs to viruses isolated over a wide time range: MABs 3/1, 5/1 and 469/6 to the WSN virus isolated in 1933 and MABs to the A/gull/Kazakhstan virus recovered in 1979. On this basis, the occurrence of the MAB 5/1 antigenic site in all of the viruses examined, including recent epidemic strains, enables this site to be considered antigenically highly conservative. The H12 clone site is somewhat less conservative; it is lacking in the viruses of two groups (4 and 7). Even less conservative are the MAB 3/1 and F-81 sites.

Of particular interest is the capacity of human epidemic viruses to bind with clone F-81. This phenomenon has been described at length by Yamnikova and co-workers (Lvov *et al.*, 1986), and the autoradiograph above provides further illustration. Accordingly, it is worth noting evidence suggesting that in an *in vitro* system the F-81 clone MABs inhibit the transcriptase activity of human but not of animal viruses (Krasko and Bukrinskaya, 1986). The relationship between the species identity of influenza viruses and the nucleoprotein has been proved by Sholtissek *et al.* (1985). It is difficult at this time to imagine the mechanisms of this relationship. It seems not unlikely, however, that the phenomenon of intracellular proteolysis of human, but not animal, influenza viruses might have a certain bearing on the mechanisms involved (Zhirnov and Bukrinskaya, 1984).

4. EFFECT OF SOME PHYSICAL FACTORS ON NP ANTIGENICITY IN A/WSN/33 (H1N1) VIRUS

The evidence presented in the preceding section points to the high variability of antigenic sites of the influenza virus NP and seems to suggest a reasonably high level of their structural organization. To assess the level of structural organization of the NP antigenic sites, use was made of a number of physical factors influencing conformation of the protein, such as heating, acid pH, and ultraviolet (UV) irradiation (Prokudina *et al.*, 1987). The ability of the viral NP to bind with MABs was analysed with the virus exposed to each of these factors. It should be emphasized that the autoradiograph presented above (Figure 1) gives grounds for believing that all the MABs utilized in this work are directed toward different antigenic sites.

Heating of proteins at + 100°C is known to cause changes in their conformation. As follows from Figure 2, the antigenic sites of the NP have proved to be exceptionally sensitive to heating at + 100°C, and

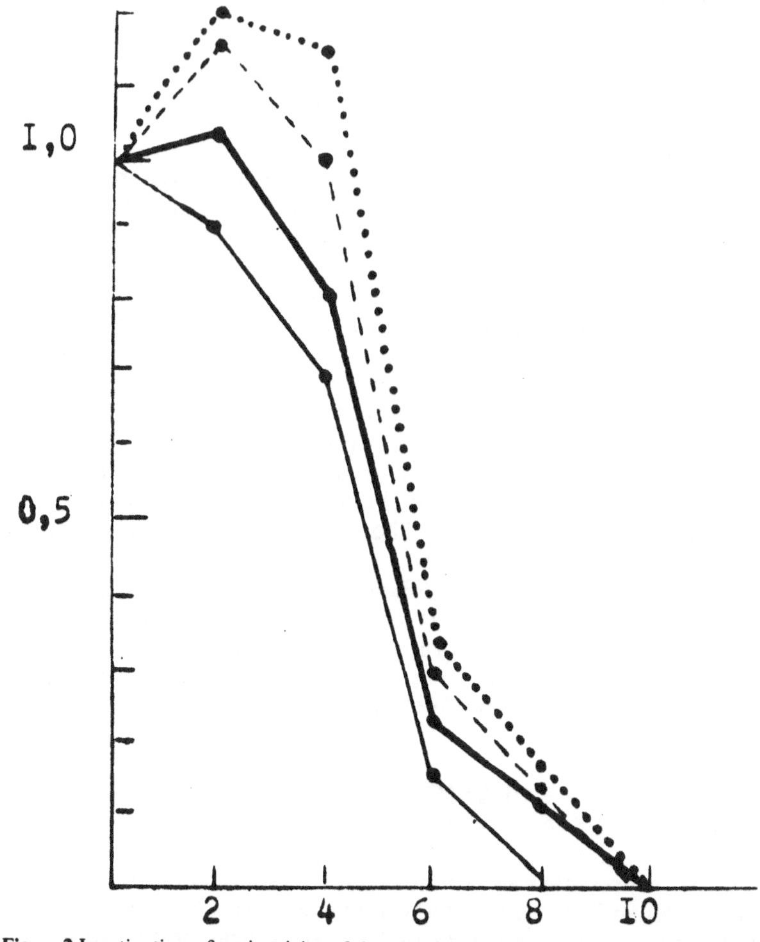

Figure 2 Inactivation of antigenicity of the NP upon heating of WSN virus at 100°C. The *abscissa* is the time (s) of virus incubation at 100°C.

The *ordinate* is $\frac{^{125}\text{I cpm at } 100° - \text{negative control cpm}}{^{125}\text{I cpm at } 20° - \text{negative control cpm}}$

The lines correspond to MAB clones: the solid heavy line, H-12; the fine line, 469/6; the broken line, 3/1; and the dotted line, 5/1.

exposure to this factor for 10 s completely abolishes the NP's capacity to bind with all MAB clones. The clone 469/1 site is the most sensitive to heating. Initial heating (for up to 4 s somewhat enhances the binding capacity of sites for clones 3/1 and 5/1, which appears to be due to initial conformational changes in the NP under these conditions, rendering these sites less accessible to the MABs. The high sensitivity of the antibody-binding capacity of the influenza virus NP to heating indicates the structural complexity of this protein's antigenic sites. As mentioned above, heating at + 100°C for 10 s completely abolishes NP antigenicity whereas, according to data in the literature, the membrane protein of the influenza virus does not lose antigenicity in spite of heating under the same conditions for 2 min (Oxford and Schild, 1976). The organization of the membrane protein antigenic sites is, therefore, much simpler, and the authors believe that this protein's antigenicity stems merely from its primary structure.

The effect of acid pH on the influenza virus NP's antigenicity was analysed by keeping the virus in buffers with varying pH, as mentioned under *Materials and methods*. The results showed that at pH 5.0 the capacity of the NP to bind with MABs was preserved, whereas beginning with pH at 4.0 and lower it decreased (not shown). The pattern of decrease of the antibody-binding capacity varied somewhat with antigenic sites, the greatest sensitivity being displayed by the site corresponding to clone 469/6. Acid pH, like heating, did not make the A/WSN/33 (H1N1) influenza virus NP acquire the capacity to bind with clone F-81.

In further experiments the virus was exposed to UV irradition. Table 1 presents rates for the binding of ^{125}I-MABs with the NP of the A/WSN/33 (H1N1) virus exposed to UV rays for variable time periods. As seen from the table, exposure of the virus to UV irradiation reduces binding rates for all MABs; again the UV-irradiated virus does not acquire the

Table 1 Effect of exposure of influenza A/WSN/33/(HlNl) virus to UV irradiation on its capacity to bind with anti-NP^{125} I-MABS.

MAB	*Without UV irradiation*	*With UV irradiation*					
		15 min		*30 min*		*45 min*	
	BR N	*BR*	*BR UV/BR N*	*BR*	*BR UV/BR N*	*BR*	*BR UV/BR N*
5/1	18	15.6	0.87	11.7	0.65	8.1	0.45
H-12	15	18	1.2	14.2	0.95	11.7	0.78
3/1	6.2	5	0.8	3.7	0.6	2.4	0.4
469/6	5.5	4.3	0.78	2.8	0.5	1.5	0.28
F-81	0.02	0.02		0.03		0.01	

BR = binding rate; N = normal

capacity to bind with clone F-81. To visualize the dynamics of loss of antibody-binding capacity by the virus as a function of UV-irradition time, Figure 3 depicts the relationships between binding rates for the irradiated and non-irradiated virus. One can see that the binding capacity of the antigenic sites corresponding to clones 5/1, 3/1 and 469/6

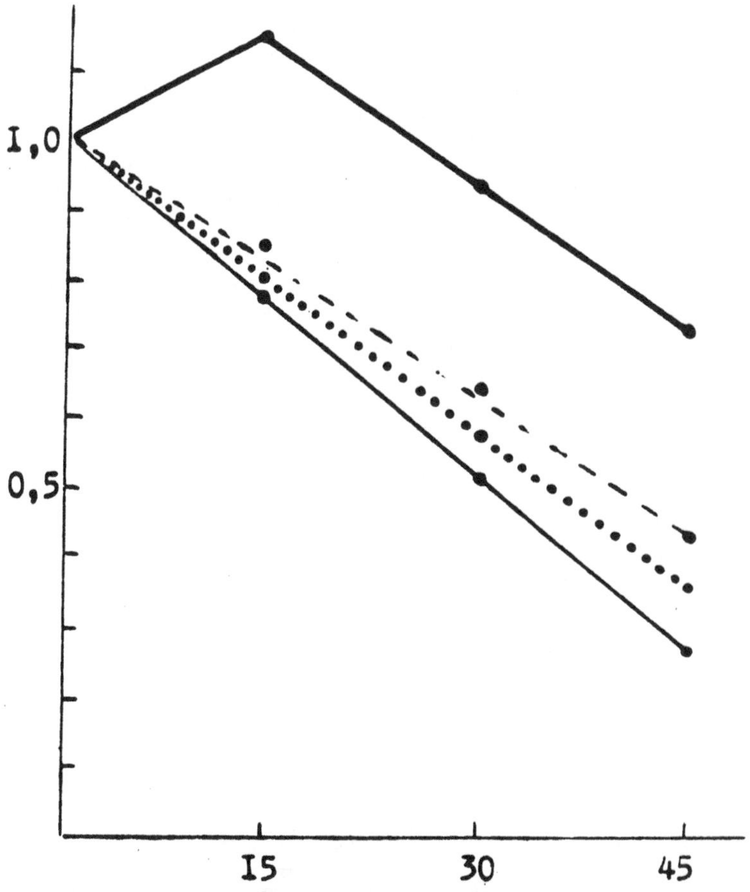

Figure 3 Inactivation of antigenicity of NP epitopes on exposure of WSN virus to UV irradiation. The *abscissa* is the time (min) of virus UV irradiation.

The *ordinate* is $\dfrac{\text{UV-WSN } ^{125}\text{I cpm} - \text{negative control cpm}}{\text{WSN } ^{125}\text{I cpm} - \text{negative control cpm}}$

The lines correspond to MAB clones: the solid heavy line, H-12; the fine line, 469/6; the broken line, 3/1; and the dotted line, 5/1.

undergoes one-hit inactivation, the site corresponding to clone 469/6 being the most UV-sensitive. The site corresponding to clone H-12 proved to be highly resistant to UV and, when exposed to UV light for 15 min, it would bind to ^{125}I-MABs even more intensively than it did when unexposed. The site may be presumed to be organized in a relatively simpler manner, or have relatively smaller dimensions, presenting a minor "target" for UV. If the presumption were valid, one would expect this site to be resistant to heating and acid pH; this is not supported by the evidence above. It should be stressed that intact virions were exposed to UV irradiation, in which the NP was known to be in intimate contact with the viral RNA. It has been reported that with ribonucleoprotein (RNP), one of the effects of UV irradiation involves covalent cross-linking at points where the NP makes contact with the RNA (Smirnov *et al.*, 1983, 1984). Smirnov and co-workers analysed the physicochemical properties of UV-irradiated RNP; their results suggest that only a limited number of RNA bases come in contact with NP molecules. Our data lead to the assumption that, in the influenza virus, NP molecules do not make equally intimate contact with viral RNA, and in particular site H-12 is relatively farther away from the RNA than other sites, which reduces its chances of being cross-linked with the RNA and provides for its relative UV resistance. UV-irradiation results appear to suggest that site H-12 may differ from other sites insofar as its structural organization or position are concerned.

5. EFFECT OF DISINTEGRATION OF ^{3}H-URIDINE AND ^{14}C-AMINO ACIDS INCORPORATED INTO THE INFLUENZA VIRUS ON NP ANTIGENICITY

It is well known that radionuclides incorporated into macromolecules exert an inactivating effect on their biological properties (Halpern and Stöcklin, 1977; Hofler and Hughes, 1969). For ^{3}H and ^{14}C the effect arises primarily from β-radiation and transmutation, as a consequence of which the nuclei of these radioactive elements convert into those of He and N, respectively. With nucleic acids, the effects noted give rise to changes in their configuration and to ruptures. The effectiveness of the damage depends on the localization of the radionuclides and decreases in the sequence DNA → RNA → protein. Structural and functional changes in macromolecules resulting from the decay of radionuclides have been reported predominantly with respect to the nucleic acids of cells and bacteriophages (Hofler and Hughes, 1969; Tessman, 1959). The literature gives us no information on the effect of radionuclides incorporated into the influenza virus on the antigenicity of its proteins. Given below are the results from investigations on the effect of decay of

^{3}H and ^{14}C incorporated into the RNA and the proteins of the influenza virus, on the antigenicity of the viral nucleoprotein.

In this series of experiments, the labelled A/WSN/33 (H1N1) virus was obtained by growing it in chorio-allantoic membrane (CAM) cultures immobilized on filter paper disks. The infected CAM cultures were incubated in the presence of 10 mCi mL − 1 ^{3}H-uridine (25 Cimmol^{-1}) or 1 mCi mL − 1 ^{14}C-hydrolysate (30 mCi per mg C atoms). The labelled virus was freed of excess isotope that had not been incorporated, for which purpose the virus-containing liquid was passed through a Sephadex G-50 column during centrifugation (Maniatis *et al.*, 1982). The unlabelled virus was subjected to a similar procedure. Both viruses thus purified showed the radioactivity of acid-insoluble material to be about 1μCi mL − 1, coming to some 85% of total radioacivity, and infectivity of 10^{7} PFU mL − 1. Thus, radioactivity per viral particle was in the neighbourhood of 0.2 disintegrations per min. The viruses were kept at +4°C, and their biological properties were analysed at selected intervals. The longest time (70 days) corresponded to an estimated 20 000 disintegrations in each viral particle (Prokundina *et al.*, 1987 a and b). The controls were unlabelled virus.

One can see from Figure 4 that with storage ^{3}H-uridine-labeled virus linearly loses its ability to produce progeny. As this takes place HA titres remain stable (not shown). The lethal effect may be due to both ^{3}H β-radiation (mean energy 5.7 keV, maximal range of several microns)and transmutation. The likely contribution of transmutation is supported by the fact that β-rays by themselves, in particular those emitted by ^{14}C-viral proteins ($^{14}C$$\beta$-ray mean energy of 50 keV, maximal range about 30 microns) cause no lethal effect (Figure 4). Since the genome of influenza virus functions as the RNP, we found it interesting to investigate in labelled viruses the state of the nucleoprotein (NP) – the protein which is in intimate contact with the viral RNA. With this in mind, we analysed the ability of the NP-labelled viruses to bind with anti-NP MABs. It was first experimentally shown that the labelled virus used adsorbs onto a polyethylene film in much the same way as the unlabelled virus does. It can be seen from Figure 5 that, with storage at +4°C, the ability of the A/WSN/33 (H1N1) virus, labelled with ^{3}H-uridine or ^{14}C-protein hydrolysate, to bind with anti-NP MABs decreases. With the ^{3}H-uridine-labelled virus (Figure 5A), differences in the degree of inactivation in the NP antigenic sites become obvious, i.e. the site corresponding to MAB clone H-12 turned out to be comparatively resistant. With either type of radioactive label, the virus did not acquire the capacity to bind with clone F-81 (not shown) the determinants for which are lacking in this virus. The reduction in antigenicity of the labelled virus NP seems to be related to the damaging action of β-rays

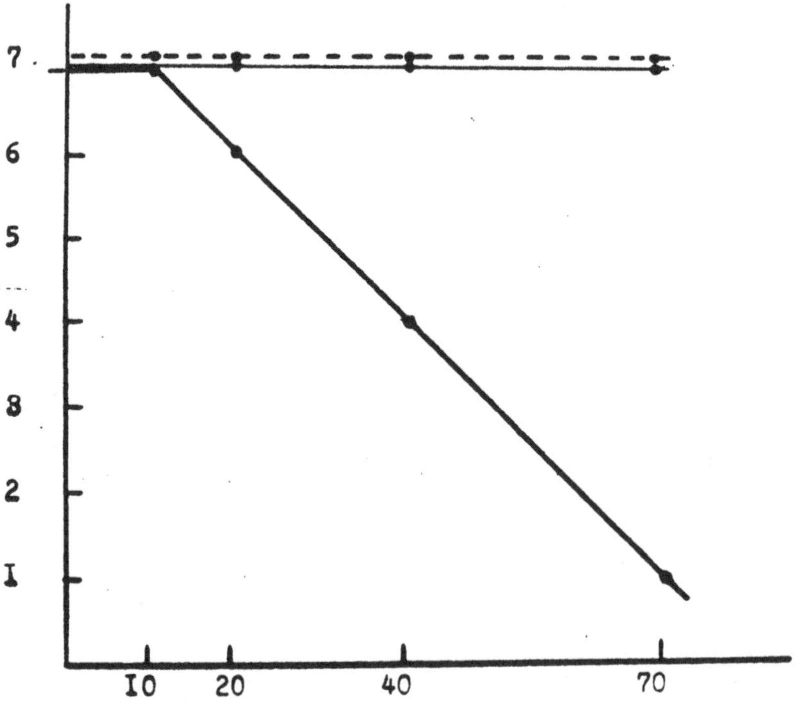

Figure 4 Effect of radioisotopes incorporated into WSN virus on its infectivity. The *abscissa* is the time (days) the virus was kept at 4°C. The *ordinate* is the virus infectivity (log PFUL mL^{-1}). Designations: the dashed line, unlabelled; the solid fine line, labelling with ^{14}C-amino acids; and the solid heavy line, ^{3}H-uridine labelling.

from radionuclides incorporated into the virus. It should be noted that the range of β-rays from ^{3}H is comparable to the influenza virion diameter, whereas that of β-rays from ^{14}C is many times greater than this diameter (Halpern and Stöcklin, 1977). β-rays from ^{14}C can therefore "permeate" not only this but also neighbouring virions. Assuming that β-rays emitted from the ^{3}H-RNA might have a direct radioactive action on NP antigenicity, one may apparently visualize the most resistant antigenic site H-12 as being spaced relatively wider from the RNA than other sites. Such an interpretation is supported by the relative resistance of site H-12 to UV irradiation causing, as is known, cross-linking of the RNA and NP at the sites of their contact (Smirnov *et al.*, 1983, 1984).

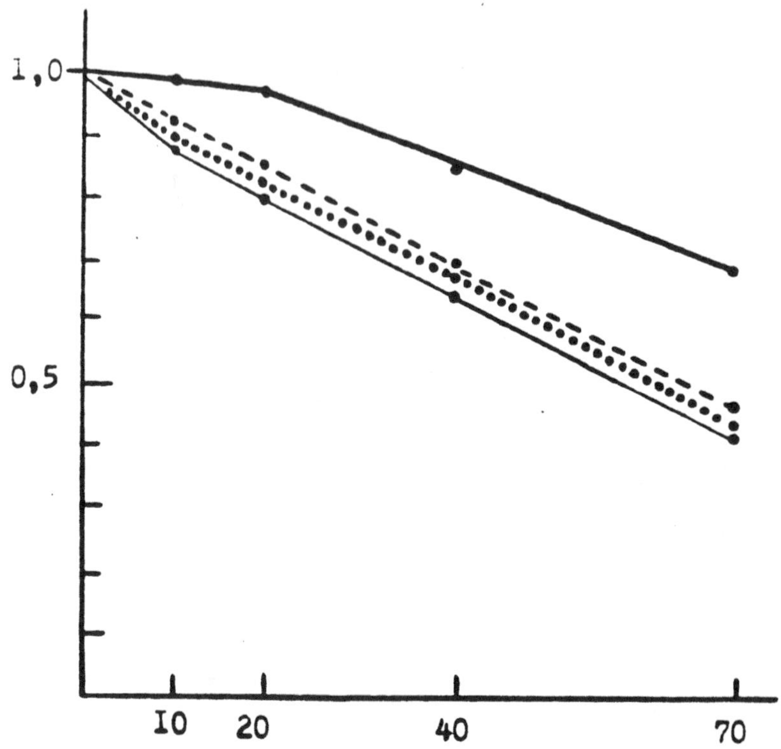

Figure 5(A)

Figure 5 Effect of (A) ^{3}H-uridine and (B) ^{14}C-amino acids incorporated into influenza virus on binding of the NP with ^{125}I-MABs. The *abscissa* is the time (days) the virus was kept at 4°C. The *ordinate* is the ratio of the degree of binding of the labelled virus to that of binding of unlabelled virus:

$$\frac{\text{labelled WSN } ^{125}\text{I cpm} - \text{control cpm}}{\text{unlabelled WSN } ^{125}\text{I cpm} - \text{control cpm}}$$

Designations for MABs: the heavy line, H-12; the dashed line, 3/1; the dotted line, 5/1; and the fine line, 469/6.

6. NP ANTIGENICITY IN RNP FRAGMENTS OF "STANDARD" AND "INCOMPLETE" INFLUENZA VIRUSES

The foregoing evidence for the influence of various physicochemical factors on the antibody-binding capacity of the influenza virus nucleoprotein indicates that this characteristic depends strongly upon

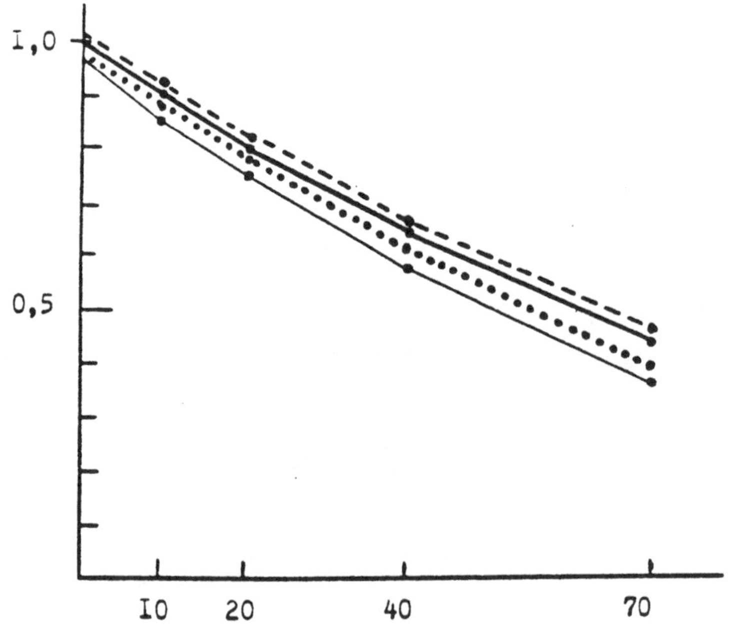

Figure 5(B)

Figure 5 Effect of (A) ^{3}H-uridine and (B) ^{14}C-amino acids incorporated into influenza virus on binding of the NP with ^{125}I-MABs. The *abscissa* is the time (days) the virus was kept at 4°C. The *ordinate* is the ratio of the degree of binding of the labelled virus to that of binding of unlabelled virus:

$$\frac{\text{labelled WSN } ^{125}\text{Icpm} - \text{control cpm}}{\text{unlabelled WSN } ^{125}\text{Icpm} - \text{control cpm}}$$

Designations for MABs: the heavy line, H-12; the dashed line, 3/1; the dotted line, 5/1; and the fine line, 469/6.

conformation of the protein's antigenic sites. Assuming that differences in the primary structure of the genomic RNA segments of influenza virus may influence the conformation of the viral RNA and, possibly, that of the corresponding NPs, an analysis was made of the antibody binding capacity of viral RNP fragments. To this end, viral RNPs were separated in a linear sucrose gradient, as described under *Materials and methods*, and selected gradient fractions were assayed. Figure 6 illustrates the distribution of the "standard" virus RNP throughout the gradient and the degree to which the NP of each fraction is bound to MABs. As illustrated, for each fraction the degree of binding decreases in the MAB sequence 5/1, H-12, 3/1 and 469/6, which is characteristic of

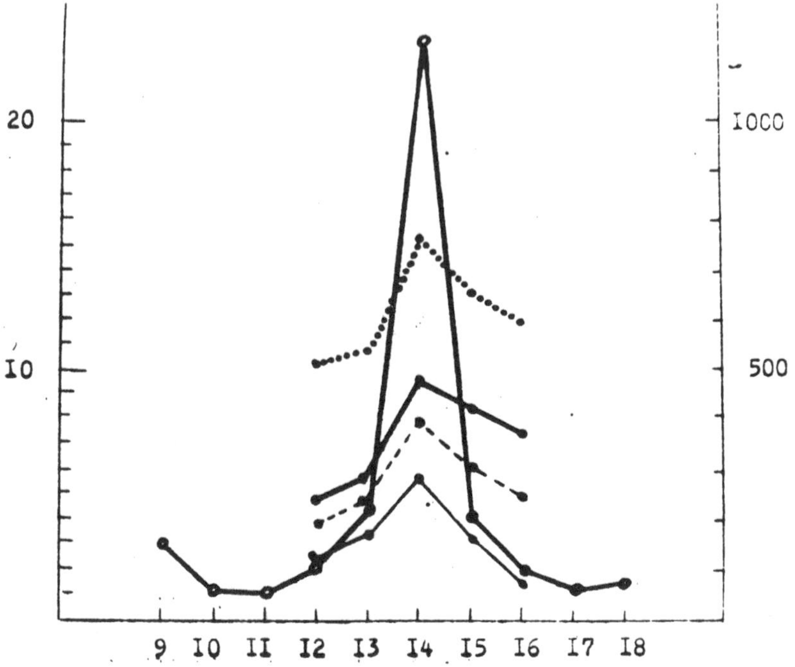

Figure 6 Binding of ^{125}I-MABs with the WSN virus RNP from different gradient fractions. The *abscissa* is fraction numbers. The *right-hand ordinate* is ^{3}H cpm of the RNA in the RNP. The *left-hand ordinate* is the binding rates:

$$\frac{\text{sample } ^{125}\text{I cpm} - \text{control cpm}}{\text{control cpm}}$$

Designations for MABs: the heavy line, H-12; the dashed line, 3/1; the dotted line, 5/1; and the fine line, 469/6.

the total A/WSN/33 (H1N1) virus (see Table 1). Importantly, all MAB clones exhibit similar dynamics of degrees of binding with respect to fractions, correlating with RNP distribution throughout the gradient. This is depicted in Figure 6 by the binding curves for the MAB clones examined being parallel. The data obtained show that similar sets of antigenic sites for the MABs examined are represented in RNPs of different size, and that these RNPs have a similar conformation. Analogous findings have been arrived at with "incomplete" virus.

It is known that after undiluted passage of influenza virus there occurs accumulation of small RNA fragments resulting from deletion of major genes (Kantorovich-Prokudina *et al.*, 1980). Accordingly, "incomplete" virus preparations contain a great many small RNPs which might presumably have original conformations and, therefore, different antigenicity compared to "normal" RNPs of "standard" virus. Given below are data describing the antigenicity of the NP of concentrated virus of undiluted passages. As can be seen from Figure 7, during the first three passages the capacity of the virus to bind with anti-NP MABs decreases, only to increase somewhat at a later stage. However, quantitative changes in binding rates, occurring with undiluted passaging, are the same for all the epitopes investigated; this manifests itself in parallel curves and demonstrates that consistent representativeness of antigenic sites in "incomplete" virus is similar to that in "standard" virus.

7. ANALYSIS OF ANTIGENIC DETERMINANTS IN THE NP OF INTRACELLULAR INFLUENZA VIRUS

This section deals with exposure of the viral NP on the surface of infected cells, and compares antigenic determinants in the NP of intracellular and mature viruses.

Our earlier data, which show that the fresh allantoic virus A/WSN/33 (H1N1) adsorbed onto polyethylene film can bind with anti-NP ^{125}I-MABs without any predestruction of the virus, are a prerequisite for accomplishing the first part of the work. Some authors are inclined to attribute such phenomena to the fact that, when adsorbed to adsorbents, the viruses undergo partial destruction, and internal determinants hidden in a native virus are uncovered in the process (Al-Kaissa and Mostratos 1982; Jaeili and van Regenmartel, 1985). Our data, however, suggest that without any predestruction or treatment, a fresh allantoic influenza virus is also capable of binding with anti-NP MABs adsorbed onto film, indicating that NP antigenic determinants are available on the surface of both the adsorbed virus and that suspended in fluid. Careful investigations by Busse and co-workers (1986), who had been looking at the nature of surface exposure of the membrane (M) protein in influenza virus, have given the authors grounds for supposing that, in addition to intact virions, partially destroyed virions in which the M protein is exposed on the surface, are also available in an allantoic virus preparation. A similar situation may also occur in respect of the nucleoprotein; however, such being the case, one should admit that in a preparation of freshly derived allantoic virus, the number of "destroyed" virions with exposed NP epitopes compares with or even exceeds that of intact virions. The nature of NP exposure, at least on some virions, the functional possibilities of such virions, and the conditions conducive to their emergence, are doubtless of interest.

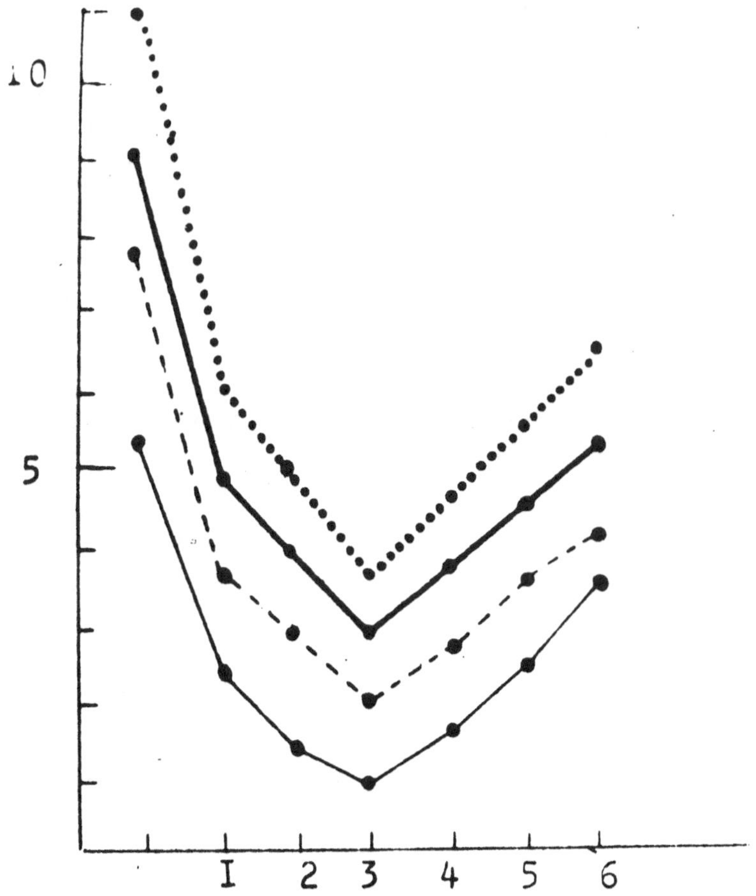

Figure 7 Binding of ^{125}I-MABs with WSN virus during undiluted passages. The *abscissa* is the numbers of undiluted passages. The *ordinate* is the binding rates:

$$\frac{\text{sample } ^{125}\text{I cpm} - \text{control cpm}}{\text{control cpm}}$$

Designations for MABs: the heavy line, H-12; the dashed line, 3/1; the dotted line, 5/1; and the fine line, 469/6.

Below we report on investigations to demonstrate NP exposure on the surface of cell-infecting virions and exposure on a cell surface of an NP synthesized *de novo* during the course of infection.

Chick fibroblast cultures (CFC) were infected with a freshly obtained virus-containing allantoic fluid (A/WSN/33 (H1N1) virus) at a calculated multiplicity of infection (m.o.i.) of 5 PFU per cell. Upon adsorption at room temperature, the virus-containing fluid was removed, and the cells were repeatedly washed with physiological saline. Next, the cultures were transferred to a thermostat at 37°C. At certain intervals a cell monolayer was further washed and overlaid with a solution containing the sum of anti-NP ^{125}I-MABs (5/1, H-12, 3/1 and 469/6). After 3 h exposure at room temperature, excess iodinated antibodies were removed, the washed cells harvested, and the radioactivity counted in a gamma spectrometer. The data presented in Figure 8 demonstrate the dynamics of the capacity of the infected cells to bind ^{125}I-MABs on their surface (solid line). As seen from the figure, the virions adsorbed to the cells maximally bind the ^{125}I-MABs immediately after adsorption. Thereafter the degree of binding decreases, evidently due to both virion desorption and penetration of viral structures into the cell. Interestingly enough, at 4h post-infection (p.i.) the capacity of the cell surface to bind anti-NP MABs increases, which must be ascribed to the emergence on the cell surface of the NPs synthesized *de novo* during the course of infection. At present we have at our disposal no data to establish whether the detected NP forms part of the virions emerging from the cells, or appears on the surface membranes as some alternative structure.

Subsequently, experiments to analyse the time course of intracellular NP synthesis were carried out. For this purpose, chick fibroblast cells infected with the A/WSN/33 (H1N1) virus at an m.o.i. of 5 PFU per cell, were labelled with ^{35}S-methionine for 20 min at 1 h intervals. The washed cells were homogenized, the homogenate centrifuged at 5000 rpm for 30 min, and the cytoplasmic supernatant applied to polyethylene film with preadsorbed MABs (summed). After 18 h of exposure the excess material under examination was removed, the film washed, and the zones corresponding to samples, computed in a β-counter. Figure 8 shows (dotted line) that at 2 h p.i. the NP begins accumulating perceptibly in a cytoplasmic fraction. It should be noted that the accumulation curve of the ^{35}S-cytoplasmic NP parallels that of the cell surface NP.

These experiments indicate that infected cells can bind with anti-NP MABs both at the adsorption stage of the infecting virus and at viral reproduction stages that may include the emergence of virus from cells. The times and dynamics of emergence of the viral NP on the cell surface correlate with those of the accumulation of the intracellular cytoplasmic NP.

Antigenicity of viral proteins is known to be governed by a number of factors affecting their conformation. In this context, viral proteins may

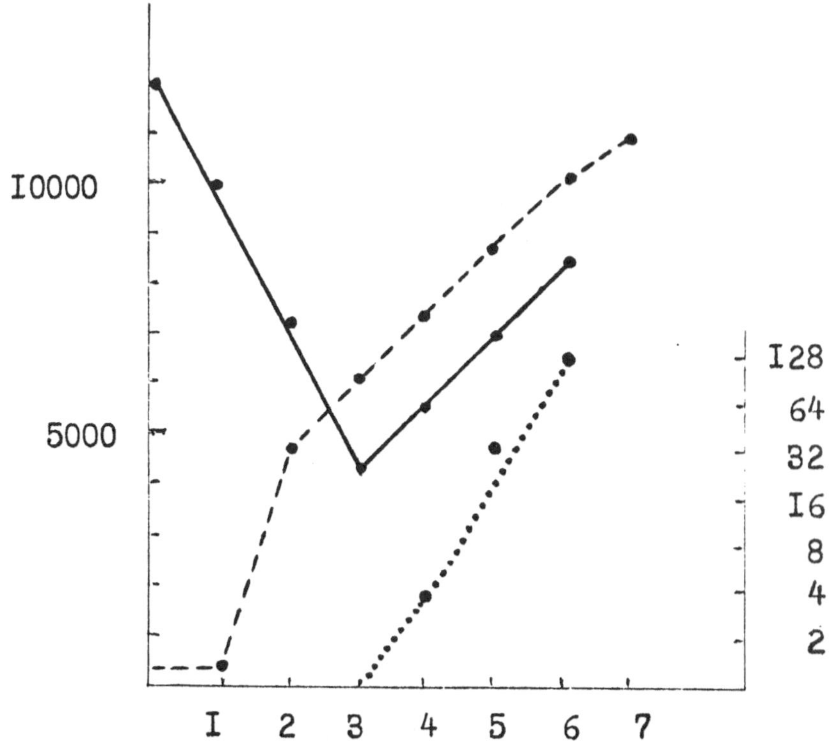

Figure 8 Dynamics of the capacity of anti-NP MABs to bind with the infected cell surface and intracellular cytoplasmic fraction. The *abscissa* is hours after infection. The *right-hand ordinate* is the HA titres in the cytoplasmic fraction of the infected cells. The *left-hand ordinate* is radioactivity (cpm). Designations: the solid line, binding of anti-NP ^{125}I-MABs (the sum total of four clones) with the infected cell surface (^{125}I cpm); the dashed line, binding of the ^{35}S-methionine-labelled fraction of the infected cells with the anti-NP MABs (the sum total of four clones) adsorbed onto film (^{35}S cpm); and the dotted line, HA.

be expected to vary in antigenicity following their intracellular processing, or intracellular viral proteins would be expected to differ in antigenicity from mature virion proteins. The literature on this subject is not abundant. But it has been reported that glycosylated and non-glycosylated forms of influenza virus HA bind with MABs in a similar fashion, i.e. they differ in neither antigenicity nor representativeness of antigenic sites in the HA (Basak and Compans, 1983). This aspect of influenza virus NP has been even less researched, although some reports suggest the intracellular processing of this protein (Zhirnov and Buknaskaya, 1984) and its phosphorylation. In the experiments outlined below we followed the accumulation of individual antigenic sites in the NP with viral reproduction and compared sets of antigenic sites available in the intracellular NP at different periods of infection with those in the NP of mature extra-cellular virus. It should be stressed that MABs to the NP of the mature extracellular A/WSN/33 (H1N1) virus were utilized in our work.

Infected CFCs were homogenized at a 1 h interval so as to produce a cytoplasmic fraction, which was adsorbed onto polyethylene film and subsequently exposed to a variety of ^{125}I-MABs. These latter were concurrently exposed to the mature extracellular virus adsorbed onto the film. The results of ^{125}I-MAB binding were registered in a gamma counter so that a binding rate (BR) could be calculated for each sample:

$$\mathrm{BR} = \frac{\text{sample } ^{125}\text{I cpm} - \text{control cpm}}{\text{control cpm}}$$

As shown above (Table 1), when different clones of ^{125}I-MABs interact with the mature extracellular virus, each clone has a different BR. This may be attributed to the varied affinity of the clones, as well as to the varied representativeness of the antigenic sites in the viral NP molecule. As seen from Table 1, on exposure to the extracellular virus the BRs of different MAB clones are in some kind of (unknown) specific correlation. Similar correlations between the BRs of different MAB clones were also preserved on binding with the intracellular NP irrespective of infection duration (not shown). Since this investigation was concerned with comparing the representativeness of antigenic sites in the NP of intra- and extracellular viruses, the correlation of the BR for the intracellular NP with the corresponding BR for the extracellular virus was calculated for each MAB clone at different points in time after infection. The results are presented in Figure 9.

It can be seen from the Figure that for each duration of infection, the BR ratios of intracellular NP to virion NP are similar for all the four MAB clones used. In this connection, there is a similar dynamics of accumulation of different antigenic sites in the NP in the course of the infection. These data suggest that the intracellular NP contains antigenic sites analagous to those in the NP of extracellular virus, their proportions being similar. It should be remarked that had we disposed of MABs to the intracellular NP, a comparison with the virion NP might have revealed a certain difference. For instance, it has been reported with respect to Sindbis virus that intracellular glycoproteins have antigenic sites masked in the makeup of the mature virion (Schmaljon *et al.*, 1983) Thus, when using MABs to the NP of an extracellular mature virus, antigenic sites similar to those present in the NP of mature virus,

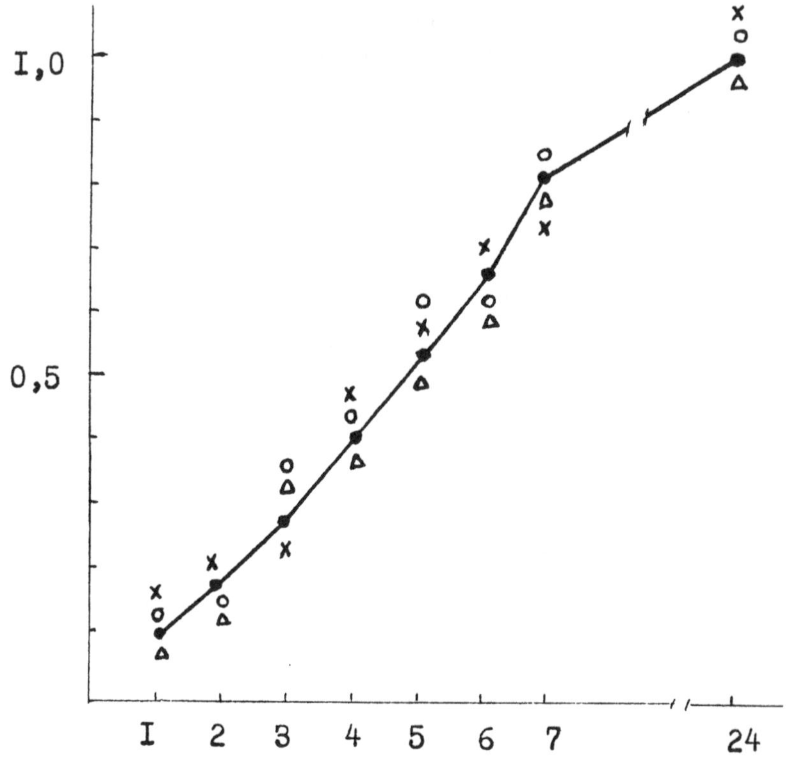

Figure 9 Dynamics of the capacity of the intracellular NP to bind with different ^{125}I anti-NP MABs as WSN virus infects cells. The *abscissa* is hours after infection. The *ordinate* is the ratio of binding of the intracellular NP to that of binding of the extracellular WSN NP:

$$\frac{\text{intracellular NP } ^{125}\text{I cpm—negative control cpm}}{\text{extracellular WSN } ^{125}\text{I cpm—negative control cpm}}$$

Designations for ^{125}MABs: ●, H – 12; ○, 5/1; × 3/1; and Δ, 469/6.

are brought out in the NP of the maturing intracellular virus in similar proportions.

In summary, the influenza virus adsorbed onto cells is capable of binding with anti-NP ^{125}I-MABs. The NP synthesized *de novo* is subsequently exposed on the surface of these cells. As virus reproduces, the intracellular viral NP shows a set of antigenic sites as representative as those occurring in the NP of the mature virus.

8. DISCUSSION

In order to analyse the antigenicity of the nucleoprotein (NP) of the influenza virus, the technique of radioimmunoassay (RIA) on polyethylene film was employed (Prokudina *et al.*, 1986 a and b; Semyonova *et al.*, 1983). Technically this method is remarkably straightforward, demonstrative, documentary, and requires tracer amounts of reagents.

In studies on the interaction between influenza virus and anti-NP MABs it was noted that WSN virus from a freshly derived allantoic fluid is capable of binding with anti-NP MABs adsorbed onto film. Pretreatment of this viral preparation with detergents did not affect its binding with the MABs. This leads to the assumption that a reasonably large number of destroyed virions with the NP protein exposed on the cell surface are present in fresh virus-containing allantoic fluid. The interpretation given fits assumptions made for the M protein of the influenza virus (Busse *et al.*, 1986). Considering, however, that the biological properties of influenza virions are fairly stable in allantoic fluid, the following questions arise. Can such "destroyed" virions be produced during the course of a natural viral infection? Can they adsorb to cells? Can they be detected immediately upon their emergence from the cells? Our experiments (Section 7) have shown that virions with the NP available to MABs can adsorb to cells. Further in the process of infection, a newly synthesized viral NP is exposed on the cell surface and it may do so in the form of maturing, initially destroyed virions (Prokudina *et al.*, 1986b)

A comparative analysis of antigenicity of the NP in a variety of influenza A virus strains confirmed this protein's marked antigenic variability, as visualized in the autoradiograph in Section 3 (Semyonova *et al.*, 1987). The presence of reasonably conservative sites in the NP for MAB clones 5/1 and H-12 enables these clones to be looked upon as promising for diagnosing influenza by RIA on film with results registered autoradiographically. The radiographed data illustrate also the validity of the assumption made by several workers that there might be a particular relationship between NP antigenicity and the species identity of influenza virus (Sholtissek *et al.*, 1985, 1987; Lvov *et al.*, 1986; Krasko and Bukrinskaya, 1986).

The high degree of antigenic variability of the viral NP has prompted the suggestion that this protein's structural organization might be complex. The high sensitivity of the antibody-binding capacity of the influenza virus NP to the action of chemical and physical factors (Section 4) lends support to the validity of this assumption (Prokudina *et al.*, 1987). Specifically, antigenic sites in the NP proved to be highly sensitive to heating at 100°C for 10 s, whereas it has been reported

(Oxford & Schild, 1976) that the reasonably simply organized M protein of the influenza virus retains antigenicity even after 3 min boiling. On exposure of influenza virus to UV light, one of the antigenic sites (for MAB clone H-12) was found to be relatively resistant to the exposure. Considering that UV-induced cross-linking is likely to occur where the viral RNA and the NP come in contact (Smirnov *et al.*, 1983, 1984), it may be assumed that site H-12 is relatively farther away from the RNA than the other sites investigated. This assumption is to a certain extent supported by the finding referred to in Section 5, namely, relative resistance of site H-12 to the decay of ^{3}H incorporated into the viral RNA (Prokudina *et al.*, 1987 a and b). The structural organization of the different antigenic sites in the influenza virus NP or their arrangement in the molecule appear to vary in some manner. Thus, evidence of a decrease in the antibody-binding capacity of the NP of the influenza virus exposed to physicochemical factors, or of the virus with incorporated radioisotopes, bears witness to the high level of this protein's structural organization and the pronounced dependence of its antigenicity on conformation.

It should be emphasized, however, that in the foregoing experiments, conformational changes in the viral NP arose from man-made impacts. Subsequently we sought to elucidate the relationship between antigenicity of the viral NP and possible conformational changes under natural conditions. These experiments were conducted with an influenza virus strain, A/WSN/33 (H1N1). We presumed, among other things, the possibility of varied conformation for the NP in different RNPs of the standard virus or in the deleted RNPs of the "incomplete" virus (Section 6). However, these experiments revealed that within a single viral strain the antibody-binding capacity of several RNPs proved to be stable, irrespective of their size and structural completeness. Next, we proposed that during maturation of influenza virus, the intracellular NP processing might produce conformational changes and, therefore, changes in its antigenicity (Semyonova *et al.*, in press). Experimental evidence indicated that, when employing MABs to mature virus, the intracellular NP exhibits antigenic sites of the mature extracellular virus, represented in a similar fashion.

To summarize, there is a fairly distinct antigenic NP variability among different influenza A virus strains and a pronounced relationship between NP antigenicity and man-made influences that alter its conformation. However, under natural conditions, within a single strain of both stationary and reproducing virus, qualitatively stable NP antigenicity is the case.

REFERENCES

Al-Kaissi, E.N. and Mostratos, A. (1982) *J. Virol. Methods*, **4**, 353–358.

Basak, S. and Compans, R. (1983) *Virology*, **128**, 77–91

Bean, W.J. (1984) In *Segmented Negative Strand Viruses*, New York, Academic Press, pp. 315–319.

Busse, T.L., Khristova, M.L., Makhov, A.M. *et al.* (1986) *Vopr. Virusol*, **5**, 544–547 (In Russian).

Compans, R.W., Content, J. and Dusberg, P.H. (1972) *J.Virol.*, **10**, 795–800.

Dimmock, N.J., Carver, A.S., and Webster, R.G. (1980) *Virology*, **103**, 350–356.

Graisler, B., Zaidel, V., Herrman, B. and Dener, L. (1986) *Vopr. Virusol.*, **5**, 538–544 (In Russian).

Halpern, A. and Stöcklin, G. (1977) *Rad. Environ. Biophys.*, **14**, 257–274.

Hofler, K.G. and Hughes. W.L. (1969) *Rad. Res.*, **47**, 904–1000.

Hunter, W. and Greenwood, L. (1962) *Nature*, **194**, 495–496.

Jaeili, M. and M.H.V. van Regenmartel (1985) *J. Virol. Methods.*, **11**, 189–198.

Kantorovich-Prokudina, E.N., Semyonova, N. Berezina, O. and Zhdanov, V.M. (1980) *J. Gen. Virol.*, **50**, 23–31.

Krasko, A.G. and Bukrinskaya, A.G. (1986) *Vopr. Virusol.*, **5**, 528–532.

Lecomte, J. and Oxford, J. (1981) *J. Gen. Virol.*, **57**, 403–406.

Lvov, D.K., Bukrinskaya, A.G., Yamnikova, S.S. *et al.* (1986) *Vopr. Virusol.*, **6**, 655–657.

Maniatis, T., Fritsch, E.E. and Sambrook, K.J. (1982) In *Molecular Cloning. A Laboratory Manual.* Cold Spring Harbor Laboratories, pp. 390–392.

Oxford, J.S. and Schild, G.C. (1976) *Virology*, 74, 394–402

Prokudina, E. N., Semyonova, N. P. and Zhdanov, V. (1986a) *J. Virol. Methods*, **13**, 27–33.

Prokudina, E. N., Semyonova, N.P., Yamnikova, S.S. and Zhdanov, V.M. (1987a) *Radiobiologiya*, **6**, 800–804 (In Russian).

Prokudina, E., Semyonova, N., Yamnikova, S. and Zhdanov, V.M. (1987b) *Arch Virol.*, 97; 325–332.

Prokudina, E.N., Semyonova, N.P. , Grigorieva, T.A. *et al.* (1986b) In *Aktual'nyye voprosy obshchei i meditsinskoi virusologii* (Current general and medical virology concerns) pp. 78–81. Moscow: Academy of Medical Sciences (In Russian).

Schild, G.C. and Pereira, H.G. (1969) *J. Gen. Virol.*, **4**, 355–363.

Schmaljon, A.L., Kokubun, K.M. and Cule, G.A. (1983) *Virology*, **130**, 144–154.

Semyolnova, N.P., Prokudina, E.N. Yamnikova, S.S. and Vikhrev, N. Ye (1987) *Mol. Genet. Mikrobiol. Virusol.* **11**, 31–33 (in Russian)

Semyonova, N.P. and Prokudina, E.N. (1983) *Vopr. Virusol.*, **6**, 747–752 (In Russian).

Semyonova, N.P., Prokudina, E.N. and Yamnikova, S.S., *Mol. Genet. Mikrobiol. Virusol.* (In press) (In Russian).

Sholtissek, C., Ronde, W., von Hoyningen, V. and Rott, R. (1987) *Virology*, **87**, 13–20.

Sholtissek, C., Burger, H., Kisner, O. and Shortiage, K. (1985) *Virology*, **147**, 287–292.

Smirnov, Y., Rodrigues-Motto, M.P. and Famadas, M.T. (1983) *J. Virol.* **45**, 1048–1055.

Smirnov, Yu.A., Kolodkina, V.L. and Kaverin, N.V. (1984) *Mol. Genet. Mikrobiol. Virusol.*, **8**, 34–36 (In Russian).

Sokolov, B.P. and Zhdanov, V.M. (1982) *J. Gen. Virol.*, **58**, 199–203.

Tessman, J. (1959) *Virology*, **1**, 263–267.

Trushinskaya, G., Ivanova, V., Russ, G. *et al.* (1985) *Acta Virol.*, **29** 428–431.

Van Wyke, K.L., Hinshaw, V.S., Bean, W.J. and Webster, R.G. (1980) *J. Virol.*, **35**, 24–30.

Zhirnov, O. and Bukrinskaya, A.G. (1984) *J. Gen. Virol.*, **65**, 1127–1134.

INDEX

Soviet Medical Reviews, Section E
Virology Reviews

Contents, Volume 1

Contents, Volume 2

Contents, Volume 3

Part 1, Etiology and Pathogenesis of Viral Infections

Part 2, Interferon and its Inducers